MICROBIAL DIVERSITY IN HONEYBEES

MICROBIAL DIVERSITY IN HONEYBEES

Charles H. Wick and David A. Wick

CRC Press
Taylor & Francis Group
Boca Raton London New York

CRC Press is an imprint of the
Taylor & Francis Group, an **informa** business

First edition published 2021
by CRC Press
6000 Broken Sound Parkway NW, Suite 300, Boca Raton, FL 33487-2742

and by CRC Press
2 Park Square, Milton Park, Abingdon, Oxon, OX14 4RN

ISBN: 978-0-367-53938-2 (hbk)
ISBN: 978-0-367-53944-3 (pbk)
ISBN: 978-1-003-08380-1 (ebk)

Typeset in Times New Roman
by MPS Limited, Dehradun

This work is dedicated to
The Grandchildren

Contents

Preface

Bacteria, fungi and viruses, oh my! The second usual comment is "those should not be there, oh my?" and likewise "well that just cannot be right." Over the past couple of years, the number and names of the microbes we have detected and identified from honeybees have grown and become a diverse collection. Microbe diversity became a common occurrence, and finally we decided to compile all this information and share. Diversity of microbes in honeybees became an assignment.

I did have an opportunity to answer one of the "oh, my" questions. I said, "well it would be easy enough not to include it (the microbe), but it does not make it go away." The answer was perfect, "ah, had not thought of that." So, when you see all these microbes being listed and some with large numbers of unique peptides, be assured that we really do live in a biological soup, a biological or rather a microbiological cloud that contains bacteria, fungi and viruses – oh my, indeed.

Living through the convergence of technologies has been exciting. What was done through long hours and tedious work in the lab in order to grow a few microbes – viruses being the tricky part of the operation – has now been reduced to a recipe for preparing samples and software. The result is the ability to identify thousands of microbes with the potential to identify tens of thousands of microbes in a short time. The tedious work has been replaced by a simple process: prep the sample, run on the mass spectrometer to determine peptides, and analyze using the ABOid software to determine all the species. I have not plated a sample for a long time; computers and software have grown up. Every analysis is a discovery.

We divided this book into practical divisions that include the principal types of microbes: bacteria, fungi and viruses. We added some specialized chapters that include detection of different species of honeybee, detection of microbes associated with the bee gut, water microbes, and the latest virus of interest – COVID-19.

We have discovered a new way to monitor the environment or to search an environmental space for microbes by using the honeybee as a microbe collector. Different regions in the country exhibit different distributions; this was particularly evident when looking at the coronaviruses. Who would have thought that these are distributed everywhere and some areas actually have forensic levels of COVID-19.

It needs to be emphasized that we have lived within this biological cloud for a long time with no obvious ill effects. Please do not do a "chicken little" and run off thinking the end of the world is coming simply because we have put names on the microbes that are around us. Now if there were a massive increase in a particular microbe, that would also be important and suggest a further look.

We view this as a start to an effort to catalog the microbes in many different venues – for example, what if we looked at food a little closer or "clean areas." Of course, a really good bio collector would be needed, and it looks like it will be a while before something is invented to replace the honeybee. The honeybee has been proven to be an exceptional collector.

Enjoy reading this book and learning something about the microbe world around us, and know that new additions to this effort will continue to increase our knowledge and serve as new information as we move ahead.

Charles Wick and David Wick

Acknowledgments

There are many to thank for their support in the honeybee work. To name a few of these hard-working people, a special thanks is extended to the enthusiastic beekeepers everywhere and their local, state and national associations, the Almond Board of California, Project Apis m., the Department of Agriculture, the University of Montana, Montana State University, and Bee Alert Technology, Inc.

We appreciate the work done by Jinnie M. Wick for research, proofreading, formatting, and organizing the manuscript into the final draft, and most importantly for her graphic design and placing all the figures into proper form.

A particularly hearty thanks to Virginia and Jaime Wick for their patience and assistance in collecting data and proofreading, while we worked all the numbers and put this book together.

Author/Editor Biographies

Dr. Charles H. Wick is a retired senior scientist from the US Army Edgewood Chemical Biological Center (ECBC) where he served both as a manager and research physical scientist and has made significant contributions to forensic science. Although his 40-year professional career has spanned both the public sector and the military, his better-known work in the area of forensic science has occurred in concert with the Department of Defense (DOD). He is a founding member of BIOid, Inc.

Dr. Wick earned four degrees from the University of Washington and worked in the private sector (civilian occupations) for twelve years, leading to a patent, numerous publications, and international recognition among his colleagues.

In 1983, Dr. Wick joined the Vulnerability/Lethality Division of the United States Army Ballistic Research Laboratory, where he quickly achieved recognition as a manager and principal investigator. It was at this point that he made one of his first major contributions to forensic science and to the field of antiterrorism; his team was the first to utilize current technology to model sub-lethal chemical, biological, and nuclear agents. This achievement was beneficial to all areas of the Department of Defense, as well as to the North Atlantic Treaty Organization (NATO), and gained Wick international acclaim as an authority on individual performance for operations conducted on a nuclear, biological, and chemical (NBC) battlefield.

During his career in the United States Army, Wick rose to the rank of lieutenant colonel in the Chemical Corps. He served as a unit commander for several rotations, a staff officer for six years (he was a division chemical staff officer for two rotations), deputy program director of Biological Defense Systems, and retired from the position of commander of the 485th Chemical Battalion in April 1999.

Dr. Wick continued to work for the DOD as a civilian at the ECBC. Two notable achievements, and one which earned him the Department of the Army Research and Development Award for Technical Excellence and a Federal Laboratory Consortium Technology Transfer Award in 2002 includes his invention of the Integrated Virus Detection System (IVDS), a fast-acting, highly portable, user-friendly, extremely accurate and efficient system for detecting the presence of viruses for the purpose of detection, screening, and characterization. The IVDS can detect the full spectrum of known, unknown, and mutated viruses. This system is compact, portable, and does not rely upon elaborate chemistry. The second, and equally award-winning, was his creation and leadership in the development of software designed for detecting and identifying microbes using mass spectrometry proteomics. Each of these projects represents determined 10-year efforts and is novel in their approaches to the detection and classification of microbes from complex matrices. Both topics are the subjects of two of his books published by CRC Press (*Identifying Microbes by Mass Spectrometry Proteomics*, 2013, and *Integrated Virus Detection*, 2014).

Throughout his career, Dr. Wick has made lasting and important contributions to forensic science and to the field of antiterrorism. Dr. Wick holds several U.S. patents in the area of microbe detection and classification. He has written more than 45 civilian and military publications and has received myriad awards and citations, including the Department of the Army Meritorious Civilian Service Medal, the Department of the Army Superior Civilian Service Award, two United States Army Achievement Medals for Civilian Service, the Commander's Award for Civilian Service, the Technical Cooperation Achievement Award and 25 other decorations and awards for military and community service.

David Wick brings 21 years of demonstrated ability to implement new or novel uses of technology. Mr. Wick is currently providing virus- and disease-monitoring services to the nation's beekeepers. Mr. Wick has introduced and briefed the novel use of Integrated Virus Detection System (IVDS) technology to Lawrence Livermore National Lab, Livermore, CA; Plum Island Animal Research Station, Orient Point, NY (detection of hoof-in-mouth recovery from barnyard ground samples); CDC Bio Terrorism Lab, Atlanta, GA; NIAID, Bethesda, MD (work on hepatitis virus); Washington Hospital Center – Washington DC (to screen patients in the ER for an early warning to a viral outbreak); PACCOM, San Diego, CA (potential use on war ships as an early warning of a viral outbreak); Titan Corp., San Diego, CA; Homeland Security Science and Technology, Washington DC (application as an air-sampling in subway stations as an early detection for a bio-attack); NYCDOPH, New York City, New York (to monitor public viral loads); Montana Fish Wildlife and Parks (bighorn sheep); Idaho Fish and Game (mallard duck die-off); University of Montana (surface recovery in several locations); Charles River Laboratory, MA (virus verification and detection in research animals); EPA – Cincinnati (waste water testing – in 2019 BVS won 2nd place in a worldwide competition to detect viruses in waste water using the IVDS technology – awarded a monetary prize).

Publication:

Fernandez de la Mora, Juan; Wick, David; Perez-Lorenzo, Luis. Singularly Narrow Viral Size and Mobility Standards from the 38.3 nm Chronic Bee Paralysis Virus and Its 17.5 nm Satellite. *Analytical Chemistry*. Submitted June, 2020. Accepted September 2020.

Abstract

Honeybees collect pollen, nectar, water and microbes such as bacteria, fungi and viruses on their foraging trips. Detecting these microbes, let alone identifying them, has been a challenge due to the limitations that constrain historically available methods. In the past, growing bacteria and fungi and then identifying and classifying them were possible only for a few species due to time-consuming methods and lack of information. With the discovery of molecular methods in the early 1990s for identifying microbes followed quickly by proteomic methods in the 2000s and advances in computers to speed up software analysis, the process of microbe detection and identification expanded, became faster and no longer requires labor-intensive efforts. This expansion of technology enabled the detection and identification of all sequenced microbes, sometimes from a single sample. As a result, during the last decade to the present we have seen the number of sequenced microbes increase from a few to hundreds of thousands. This rapid growth in information and the parallel advances in hardware and software have enabled a more detailed look at the vast variety of microbes and in particular those found on the honeybee. More precisely, this scientific growth has allowed the development and use of mass spectrometry proteomics (MSP/ABOid) to detect and identify microbes and is limited only by the number of microbes that are sequenced. As of 2020 there are over 240,000 bacteria sequenced, of which 17,700 are completely sequenced; 38,000 viruses sequenced with over 35,000 completely sequenced as well as over 6,000 fungi with 80 completely sequenced, according to the National Center for Biotechnology Information. There is a potential of more than 52,000 completely sequenced microbes for detection and identification use. This book uses a portion of this potential to search for microbes in samples of honeybees collected all over the United States. The findings are presented in chapters according to functional groups including microbes associated with the bee gut, bacteria, viruses, fungi and others. A chapter on COVID-19 is included. Results show a rich and diverse assemblage of microbes collected by these busy bees.

List of Abbreviations

ABOid	Agents of Biological Origin Identification
IVDS	integrated virus detection system
MSP	mass spectrometry proteomics
NCBI	National Center for Biotechnology Information
SARS	severe acute respiratory syndrome coronavirus 2 (COVID-19)

1 Microbe Diversity in Honeybees

Pollinators compose two main groups: commercial beekeepers and hobbyists. Naturally, there is gradation between these two groups – some hobbyists have extensive operations that start to look like commercial operations. Pollinators provide bees to pollinate important food sources; among the largest of these efforts is the pollination of the almond crop in California in the spring. Most of the pollinators are trucked to California for this purpose. Once finished with the almonds, they are moved to other areas to pollinate apples and other fruit crops and then on to the next crop that is in bloom. Moving the bees from one site or region to another has been done for many years, and beekeepers can be thanked for this important contribution to agriculture and the food we enjoy. A question or at least a consideration has been "what is the microflora environment where the bees are working or what is the microflora where they will be working?" A possible concern is that the bees might pick up an unwanted microbe along the way. Likewise, it would be good to know if the bees picked up an unwanted microbe from one region before traveling to another. This reference book should help provide a microbial baseline from which to compare one region to another and one honeybee group with another.

Honey producers and packers face similar questions about the normal microbe environment. It could be expected that honey which comes from bees would reflect the naturally found microbes in their product; indeed the unique flavor of some honey may reflect the unique background associated with honey production. With nearly 40 million pounds of honey produced each year, the question of naturally occurring microbes, regions, and national averages for microbes becomes important. Particularly in light that often the demand for honey in the United States is greater than the supply. Honey imported from outside can be expected to have evidence of a different microflora and this information could be useful in checking the origin and contents labels.

The book in organized into chapters that represent the major microbes and groups of microbes collected by honeybees and should provide a useful reference to beekeepers, commercial beekeepers and hobbyists alike to provide the microflora, the natural microflora, that honeybees collect while working.

Foraging honeybees collect nectar and pollen and a wide variety of microbes. This book focuses on these microbes which can be divided into three main groups – bacteria, viruses and fungi. Although this grouping leaves out other thingssuch as organelles and plasmids, we can be sure that they too are being collected by the honeybees and will be analyzed and included in the listing of microbes as technology and interest permit. One very nice feature of the mass spectrometry proteomics/ABOid (MSP/ABOid) method of detection and identification is that it is not necessary to resample since the results are saved as a computer file, and the file needs to be re-analyzed at the appropriate time when the software is updated as discussed in Chapter 2 (Mass Spectrometry Proteomics/ABOid).

The MSP/ABOidmethod of microbe detection and identification utilize the genomic information available in the DNA or RNA of each microbe. This information is simplified by determining the sequence of the nucleotides in each microbe and using this information in turn to classify and identify each microbe. Nearly 254,000 prokaryotes (bacteria), 41,500 viruses, 12,100 eukaryotes (fungi, plants), 16,500 organelles and 22,000 plasmids are sequenced. The current list of sequenced microbes is available from the U.S. National Library of Medicine.

It is important to know that some of these sequenced microbes are not fully sequenced or complete. This missing sequence information may be important to make an identification. As a result, where possible, only complete sequences are used. This reduces the number microbes available for identification, as there are presently ~18,000 bacteria, ~72 fungi and ~36,000 viruses which are fully or completely sequenced. This gives us 54,072 microbes to utilize. Data groups used in this book represent a portion of these microbes. It is also important to know that as new sequences are completed, they can be easily added to existing data groups and existing data files reexamined producing "new" results from old files.

The structure of this book is divided according to areas of interest. The first chapter discusses the honeybee. Without this discussion we would not know the unique place that this insect plays in the collections of microbes. The chapter continues with a discussion of exactly what is a microbe and how are they classified or sorted so they can be distinguished from each other, and a brief discussion on how the honeybee collects. This discussion helps us understand the honeybees' collection of microbes.

How we detect and identify microbes is discussed in Chapter 2. Utilizing the Mass Spectrometry Proteomic/ABOid method is presented and why it was selected as the identification method of choice as well as how it works is discussed.

Chapter 3 discusses the honeybee (*Apis*). This chapter also has information on the detection and identification of three other species of sequenced honeybees of the genus *Apis*.

Each of following chapters presents the various groups of microbes as they occur in nature and collected by honeybees. The national average was determined as well as averages for five regions: California region, Florida region, Idaho region, Iowa region and the Montana region. Chapters 4 (Bee Gut Microbes), 5 (Coronaviruses), 6 (Bacteria), 7 (Fungi), 8 (*Nosema*), 9 (Viruses) and 10 (Water) are about the different microbe groups collected by the honeybees.

Other questions are also important in understanding this collection of microbe information, and they are addressed first, such as 'What are microbes and where do we find them?'

1.1 ABOUT HONEYBEES

In the United States, the species of honeybee that is universally managed is the Western honeybee *Apis mellifera*. The general organization of the honeybee hive is a queen bee, a large number of female worker bees (40,000–100,000) and a few drones depending on the strength of the colony.

A bee colony follows an annual cycle that begins in the spring and ends in the winter. It starts with a rapid expansion of the brood as soon as pollen is available for feeding the larvae. Breeding accelerates towards May producing an abundance of harvesting bees that are synced with the nectar. There is some variation among regions and some variation among commercial hives as the nectar cycles vary.

It is the activity in the spring that gains attention as to collection of microbes. As the climate warms and is moist, the population of microbes can be expected to increase accordingly. It would be interesting to follow the microbe population over a few seasons and years to develop a pattern of microbes seen in bee populations.

1.2 WHAT IS A MICROBE?

In the "Tree of Life" there are three principal branches, namely eukaryotes, prokaryotes, and viruses. Two of these (prokaryotes and viruses) are small organisms and generally not visible to the naked eye. These are referred to as *microorganisms*, or *microbes*. They are the microscopic organisms and exist as either single-celled or as a colony of cells. They may have been the first forms of life on Earth.

Ancients, as long ago as the sixth-century BC, suspected microbial life but had no direct way to observe it. The invention of the microscope by Antonie van Leeuwenhoek in 1670 changed this, and with the invention of the electron microscope in the 1930s it was possible to observe both bacteria and viruses.

Microorganisms include all unicellular organisms and so are extremely diverse – there may be trillions of different kinds of microbes. They occupy a niche and live in almost every known habitat. They are known to work together as symbiotes, where a bacterium can provide enzymes to a plant or a fungus and have other complex associations found everywhere.

Microbes can be both helpful and hurtful causing fermentation for good uses and plagues (bad uses) in about equal abundance. Microbes are important and it is essential that they are studied, and more than that it is essential that we are able to detect and identify them.

1.2.1 Bacteria

Bacteria have been around for a long time and may represent some of the first organized life on Earth. Unlike the cells of animals, fungi and other eukaryotes, bacteria cells do not contain a nucleus or for that matter other "membrane-bound" organelles and belong to the prokaryotes. Following the discovery of

DNA/RNA and the resulting ability to group or classify living things by their relationships due to similar DNA/RNA, bacteria were grouped and then regrouped as technology improved. Today, prokaryotes are divided into two kingdoms: the bacteria and the archaea. The kingdom of bacteria contains over 260,000 species that have been sequenced. The kingdom of archaea has over 5,000 species that have been sequenced.

Bacteria are widely or rather vastly distributed throughout the world and can be found in nearly every habitat. They are larger than the viruses and smaller than the fungi. Typically, a bacterium can be expected to be in the one- or two-micron range in size and can have many shapes. Considering that there are millions of bacteria cells in a gram of soil or a milliliter of fresh water, it is not surprising to find bacteria collected on honeybees. The only question is how many different types of bacteria are found and what are their names.

Chapter 6 includes details of the bacteria that have been isolated, detected and identified from samples of honeybees.

1.2.2 Fungi

Fungi belong to the kingdom of fungi, which are part of the eukaryotes. Fungi have been around a long time probably dating back 400–500 million years or longer. They are diverse and can be found, much like the bacteria, in every habitual niche. The two phyla of the most interest are the Ascomycota and the Basidiomycota, which are contained within a subkingdom Dikarya, since they may contain the most abundant species of the more than several millions of species of fungi.

Fungi are familiar to most everyone as they are seen as the mushrooms, the fuzz on food and seen in the environment growing on nearly everything. Although they are often thought of as inconspicuous until seen, they are important as major decomposers, producers of antibiotics and in making food.

The Ascomycota are most likely the most abundant fungi in the environment, and as they have both a sexual state and produce ascospores and an asexual stage and produce conidia, they fill the environment with spores. When spores are combined with hypha and other fungi debris, our natural habitat is covered. It is likely it is this material and whatever can be picked up from the Basidiomycota and other fungi the honeybees are collecting. One of the largest and most familiar environmental ascomycota are the *Aspergillus* species, which includes *Aspergillus niger* which is the source of the black mold found on most things, including bread.

The fungi are reported in Chapter 7, where the most common fungi are discussed as well as a national average for the more than 200 fungi collected, detected and identified from the honeybee. A detailed list is included in Appendix G.

1.2.3 Viruses

Viruses are smaller than bacteria generally ranging in size from 20 nm to 300 nm. A nanometer is one billionth of a meter. It was not until the 1930s with the invention of the electron microscope that viruses could be imaged. Viruses are referred to as *submicroscopic organisms* and are known for the feature of replicating only inside other living cells. There has been a debate over the status of "living", but that is not our concern. We are concerned about where they are found and what is their distribution.

Like bacteria, viruses have been around for a long time and may have been important during the formation of the first life on Earth. It is expected that there are millions and millions of viruses in the environment, maybe even trillions of different species considering their uncanny ability to mutate almost at will as they "react" or accommodate to their environment. It is further expected that viruses inhabit nearly every environmental niche in the world.

Presently there are over 40,000 viruses sequenced and available for use. This has been helpful in classifying and sorting out the relationships between different viruses. This process continues, and many viruses are now named and can be identified by using their sequence as a basis.

Viruses can be expected to be picked up by the honeybee, either directly or indirectly. For honeybees not to pick up a virus would be strange indeed. The question is then "how many different strains and where are they distributed much like the bacteria".

1.3 SOURCES OF MICROBES

The honeybee has many sources in which to pick up a microbe or two. Plants harbor bacteria and fungi and even viruses. Water sources are abundant and water is needed by the honeybees and they can be expected to stop at all sorts of water supplies. Many environmental sources of microbes are available to the honeybee such as farms, other cultivated sites, urban sites, general rural areas, forests and numerous other areas. Other opportunities are also available as the natural microflora is disturbed by weather and weather events. These topics are discussed further.

1.3.1 Plants

Plants both benefit and suffer from microbes. There are many microbes, bacteria and fungi that help plants by providing an abundance of nutrients and minerals by breaking down organic matter. Others benefit the plants by actually making a symbiotic relationship where the microbes fix nitrogen to facilitate water and nutrient uptake and provide sugars, amino acids and other nutrients to the plant. On the negative side, there are many microbes that attack and cause a variety of diseases to cereals, corn, tobacco, tomato and nearly all of our food-producing plants. Likewise, microbes attack trees and other plants. All part of the cycle of breaking down complicated organic matter into simple compounds so they can be used by the complicated organic organisms. It is this process that produces a biological fog around most habitats.

Plants are in the thick of the biological fog, a fog full of bacteria, fungi and viruses and their debris from all the activity. The busy microbes are also in their own cycle of life producing more bacteria, fungi and viruses that all add to the fog. It is within this fog that the honeybee forages, and in the process of interacting with flowers and other plants in the environment pick up the bacteria, fungi and viruses.

Looking through the list of bacteria in Chapter 6, the fungi in Chapter 7 and the viruses in Chapter 9, the relationship between plants and microbes is evident by seeing the many mycorrhizal microbes and microbes associated with reducing complex compounds and in this way adding to the nutrient cycle that benefit the plants and even other microbes. Some of these interactions involve complex relationships.

1.3.2 Water

If there is water it is most likely there will be microbes. Many of these microbes are just doing their task of reducing complicated organics into simplified organics and recycling nutrients. Some of these microbes are innocuous and others cause disease. Most of the disease-causing microbes in water are bacteria. This is the main reason we have water treatment facilities – to prepare our water (remove microbes) for drinking. Since water contributes to the biological fog that surrounds our habitats and is essential for life processes, it makes abundant sense that the honeybee would be collecting water and along with water, whatever microbes are present.

Different types of water may influence the type of microbes collected by honeybees. Free-flowing water such as ponds, puddles following rain, stagnate water, and other sources all provide a source for the honeybees. The area surrounding the working area of a hive may have multiple sources of water, all of which may have a different microflora available to the honeybee.

Chapter 10 details those microbes associated with water purity testing. It is interesting to see what microbes are listed and learn about their environmental niche.

1.3.3 Environmental

The natural environment has an impact on the type of microbe that the honeybee may pick up. There are many reaching from the North Pole to the South Pole with all the differences in-between. The most microbe-rich areas are those where it is warm and moist – tropics and subtropics. In this region the plants and animals provide ample material for the microbes to recycle. Likewise, in the sea and water areas, the microbes seek to recycle nutrients. Even clear water lakes have their microbe inhabitants. Arid places have their own microbe residents, and some fungi are perfectly happy growing in the desert. Usually just not in the abundance seen in warm and moist areas.

Every domicile has a host of microbes in residence which most likely fluctuate during the seasons. Weather disturbances, such as hurricane, can cause a change in the environment and a change in the microbes. Not so much a change in the species, but a change in the population density as the microbes multiply.

Forested regions and similar spaces have microbe populations that vary. Orchards and cultivated areas have their own unique population of microbes. This population is influenced by the use of fungicides, insecticides, and other treatments. Nevertheless, there is a microbe population in residence in all of these various environments.

The honeybee collects microbes from all environments, and the resulting collection may reflect this work. It is evident that the normal microflora around the country is similar, and the difference between regions is more about the population density of the microbe rather than the name of the microbe.

A special environment, an urban environment or a specialized area such as a hospital, may represent a unique opportunity. The honeybees will pick up the microbes associated with such an area in just as fastidious a fashion as any with other area. See Chapter 5 that illustrates the collection of COVID-19.

Each chapter and each microbe collection illustrate the diversity of the microbes found and collected.

1.3.4 Other Sources

Waste sites and special sites where human activity or natural activity has changed an environment can be another source of microbes for honeybees. These areas provide very different conditions and in some ways are an artificial site. Microbe populations of all sorts may be different. Nevertheless, the honeybee if working close to such a site is likely to sample from it and carry back to the hive the bounty of the site. This may provide an opportunity to monitor these sites and follow the microbe population over seasons or years to understand or at least assess any danger to other environments or to plants and animals.

1.4 FLIGHT OF THE HONEYBEE

Life in a honeybee (*Apis mellifera*) colony is perennial. The average lifespan of a queen is three to four years and the workers a few weeks to several months, depending on the climate. The worker bees live in large colonies of up to 100,000 individuals. It is the workers that perform the "flight of the bumble bee" as they forage to find a nectar or pollen source, bring that information back to the hive and then go to work harvesting.

2 Mass Spectrometry Proteomics

A central advantage of MSP is the ability to add newly sequenced microbes to a data group. A new sequence is downloaded from the National Center for Biotechnology Information (NCBI), processed and added to an existing data group. This updated data group can then be used to reexamine a RAW file to determine if the new microbe is detected and identified. Molecular methods are generally limited to the detection of a particular microbe or a group of microbes (microarray). These molecular methods can be used to verify a detection, but when time is not a factor and a wide-ranging group of microbes are involved, it is easier to use the MSP. Further, it is a time-consuming process to add a newly sequenced microbe or indeed a group of 30 or more newly sequenced microbes to molecular-based systems. Physical methods, such as the Integrated Virus Detection System (IVDS) (Wick, 2015), do not have this limitation; since it is a physical counting methodology, any new virus will simply be counted and sized. Nothing special in being able to detect a new virus. Although it can detect viruses quickly, it does not directly identify them; therefore once a virus is detected, a confirming method is usually used to confirm the identity. Integrated Virus Detection System (IVDS) is used to detect viruses. Since bacteria and fungi are much larger than viruses, IVDS does not directly detect them, but may detect viruses associated with them and specific bacterial components of virus size (US Patent No. 7,850,908 B1, Wick, 2010). A proper use of IVDS is as a front-line screening device to detect virus particles in a sample or more importantly to determine if there are no virus particles in a sample (presently being used to screen for COVID-19). The latter capability reduces the number of samples being analyzed by other means that are either expensive per sample or take a longer time. IVDS does an excellent job detecting all the viruses in a single sample and that is useful for detecting an unknown virus (unknown) or a virus that has mutated or otherwise changed. It can do this since it is a physical method and simply counts all the virus particles in a sample and separates them by their size. MSP allows for the detection and the identification of all the microbes in one sample. The accuracy of this identification is based on the microbe sequence, same as molecular methods, and is thus able to identify to strain where available.

2.1 MSP METHOD

The concept of operation is to collect a sample, prepare it for the mass spectrometer, have the mass spectrometer create a RAW (unprocessed data, it is a file extension) file, analyze the RAW file to determine the peptides detected and use these peptides to determine which ones are unique to the various microbes. Thus, each microbe detected is done so by determining a match from the environmental sample peptides with those unique peptides, determined by calculation based on the sequence of the microbe. In this manner all the sequenced microbes are associated with their unique peptides and matched with the sample peptides collected from the environment – in this case the honeybees.

The physical process is: samples of honeybees are collected from various sources and processed for the MS according to the following steps: step one, make bee smoothie, step two, follow a sample preparation recipe, run though MS, analyze the RAW file produced by the MS using the ABOid software and then make a report.

Agents of Biological Origin Identification (ABOid) is an analytical tool invented and patented by the US Army (US Patent No. 8,224,581 B1, 2012; US Patent No. 8,412,464 B1, 2013) and under license to BIOid, Inc. ABOid is fully described in Chapter 6 of a book by Wick (2014, *Identifying Microbes by Mass Spectrometry Proteomics*). ABOid identifies a microbe by searching the sample ion spectra of peptide ions against theoretical peptides determined by calculation from microbe DNA/RNA sequences. ABOid simply determines those unique peptides for each microbe and compares the unique peptides with those peptides identified in a sample. The result is a highly accurate, gene-based identification of viruses, bacteria and fungi among other sequenced organisms downloaded from the National Library of Medicine, the NCBI. ABOid is thus the tool of choice when seeking to identify large numbers of microbes in a sample. In the case of COVID-19, all the sequences of the various strains were downloaded from the NCBI and added to a data group and used to identify COVID-19 in a collected sample.

ABOid has been used to analyze thousands of environmental samples and has identified nearly a thousand microbes. This is particularly important when considering the number of microbes to be identified and the limitations of many historical methods the challenge is to include as many of the available microbes as possible: at the time of writing this book, there are about 254,000 prokaryotes (bacteria), 38,500 viruses and 12,100 eukaryotes (fungi, plants) that are sequenced. There are also 16,500 organelles and 22,000 plasmids. To make a new primer or spectra or antibody or similar detection method simply takes too long and costs too much money. These historic methods struggle to keep up, but can be used later when there is time to develop the special needs and control the costs.

Given the success of ABOid in identifying and classifying microbes in complex environmental situations and the ease of sampling, ABOid is a compelling tool to use anywhere there is a need for microbe identification that includes bacteria, viruses and fungi. Since this can be accomplished in one sample, one effort and at low cost with an archival record, ABOid has been selected as the analytical tool of choice.

It needs to be noted that once a sample is run though the MS and a RAW file is generated, the report can be generated showing the microbes in the sample. If new microbes are added to a data group, the existing RAW file can be re-analyzed to determine if the new microbe or microbes are present in the sample. It was an easy step to examine some of the historical RAW files using the updated data groups to determine if a microbe was detected and identified. An example of this step will be found in the chapter on coronavirus (Chapter 5) where coronaviruses were downloaded from the NBCI and put into their own data group. Some of these sequences were only weeks old when used to examine honeybee samples collected before and during the COVID-19 outbreak.

2.2 DEFINING DETECTION AND IDENTIFICATION USING MSP

A convergence of hardware (mass spectrometers) and software (ABOid) and faster computers and rapid additions to sequencing methods all contributed to the success of MSP/ABOid. The first success was for single microbes. Grown in a microbiology laboratory, bacteria were identified, collected and prepared for the MS. Individual MS files (RAW files) were then analyzed using the ABOid software, and the bacteria was confirmed. Additional efforts combined bacteria and repeated the procedure which then detected and identified multiple bacteria (Jabbour et al., 2010) (Wick, 2014, *Identifying Microbes by Mass Spectrometry Proteomics*). Details and examples of MSP/ABOid detection and identification of COVID-19 are included in Chapter 5.

2.3 ADDING MICROBES

Figure 2.1 shows five microbes. These microbes represent the microbes in an original data group. They were identified and their sequences were downloaded from the NCBI and added to the original data group. A sample file (RAW) was processed through the MSP is then analyzed using the ABOid software.

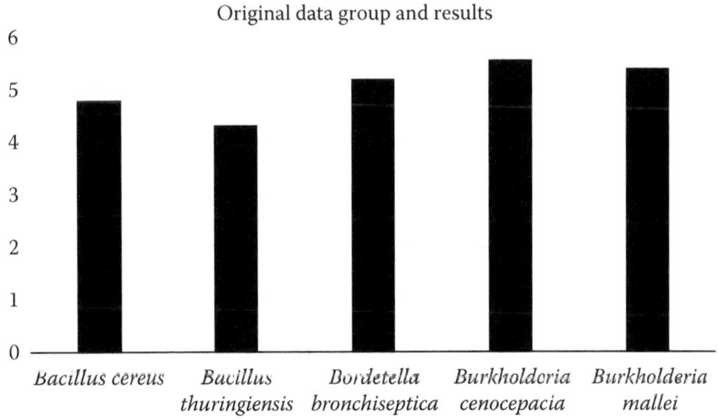

FIGURE 2.1 Example of how to add microbes to a data group. These five microbes represent an original data group that was analyzed by the MSP/ABOid method, and their average numbers of unique peptides are presented in relation to each other.

Figure 2.1 is the result. Five other (new) microbes were identified as being of interest and their sequences downloaded from the NCBI and added to the original data group. The new data group now contains the original five sequences and the new five sequences for a total of ten microbes. Figure 2.2 shows the 10 new microbes and a re-analyzed result which contains the new microbes and the new results that demonstrate the process. In this manner numerous new microbes can be added as their sequences become available.

New microbe sequences are being added every day to the NCBI database. Considering that the number of sequenced microbes has increased from a few hundreds to more tens of thousands, it can be seen that it is useful to add these new microbes to any detection and identification method.

The original virus data group used in the analysis of honeybee samples consisted of less than 100 viruses. This number was increased to 200, 300 and then 750 viruses by simply downloading new virus sequences and adding them to the virus data group. This process can be accomplished in less than an hour moving at a deliberate and scientific speed, which makes the new data group more robust and useful. When the first 200 new virus sequences were added to the virus data group, it was interesting to see the new viruses show up in the MSP/ABOid analysis. The old charts could be compared with the new charts, and it was straightforward to pick out the new detections and identifications. Sometimes there were several new viruses.

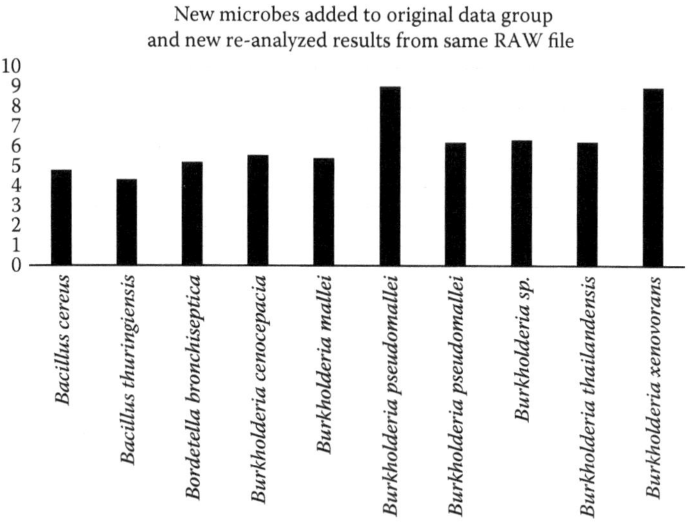

FIGURE 2.2 Example of how to add microbes to a data group. In this case, five new microbes were added to the original data group seen in Figure 2.1. The original RAW file is re-analyzed to produce the new results and for the detection and identification of the new microbes. This represents the re-analysis of an existing data file and demonstrates the ability to analyze archived data with new microbe data without re-sampling and just updating the software.

Bacteriophage sequences were added to the virus data group, and the result was the ability to detect and identify many bacteriophages. This was important as a means to indirectly detect the associated bacteria from plants such as tomato and citrus and other plants.

Likewise, microbes loaded into different data groups can be compared to look for those associations between fungi and bacteria, bacteria and viruses, and so on.

This capability was exercised in looking at the new COVID-19 outbreak. When the sequence of the novel SARS virus isolated from Wuhan-Hu in China was available, it was downloaded from the NCBI and a new data group was created. It was not long afterwards that additional sequences were made, and these were downloaded and added to the data group, and likewise until more than a hundred sequence of the COVID-19 were added. More details on COVID-19 are given in Chapter 5.

Adding bacteria and fungi is accomplished in the same manner. It is also possible to create specialized data groups for particular groups of microbes and organisms such as those discussed in Chapter 3 – *Apis*, Chapter 4 – Bee Gut, Chapter 5 – Coronavirus, Chapter 6 – Bacteria, Chapter 7 – Fungi, Chapter 8 – *Nosema*, Chapter 9 – Viruses and Chapter 10 – Water Microbes as sequences become available for use. Hundreds of eukaryotes, prokaryotes and viruses have been added to the current data groups used for analyzing the microbes carried by honeybees.

In this manner selections from the over than 10,000 eukaryotes, 250,000 prokaryotes and 40,000 viruses can be added until they are all added. This would make a very robust capability to detect microorganisms.

Bacteria and fungi have had other methods for determining a classification scheme without the use of a genetic sequence. Current methods, however, have mostly replaced these classical schemes with the new one based on DNA or RNA – their genetic sequence. In a similar manner, microscopy can detect microbes (bacteria and fungi), and chemical methods such as Gram stain, fermentation, grown on different media and colony characteristics have all been used to classify bacteria and fungi. Methods based on the genetic sequences of these microbes have allowed a more detailed understanding of their phylogenic relationships and a more rigorous means for determining an identification. Identification of microbes by using their genetic sequences has also allowed detection and identification by electronic means, such as the MSP/ABOid method which is based on molecular analysis and uses computer software to identify.

Likewise, viruses are classified by their genetic sequences. Particle counting methods such as the IVDS uses a physical ion-mobility method to count the individual viruses and separate them by their size (Wick, 2015). By separating viruses by size, IVDS has the ability to give an approximation of the type of virus, usually classifying them to family groups. The major benefit of IVDS is that it is a rapid means to detect the presence of a virus in a sample and by using size to classify it to a preliminary identification, for example, influenza is 92 nm or 102 nm in size depending if it is type A or type B. Another benefit of IVDS is that since it is not restricted by chemical reactions, it can detect all the viruses in a sample at the same time. This feature makes IVDS a perfect method to screen samples and verify identification by another means, such as MSP/ABOid. Since most samples from the environment are typically negative (unless looking at a preselected population) for flu or COVID-19, IVDS is again indicated as a means to screen large numbers and only referring positive samples for verification.

All these earlier methods are useful, but it has become the standard to verify a detection based on a classification scheme based on DNA or RNA sequences. The resulting phylogenetic relationships are useful in following genetic drift and mutations in a particular strain that result in a new strain for a particular microbe. This ability to mutate and create new strains of a microbe is responsible, in part, for many of the new sequences that we have seen over the past ten years.

Other reasons that contribute to the increase in the number of microbes sequenced are that there are probably trillions of microbes and the development of fast-sequencing instruments. These two reasons point to a method such as MSP/ABOID for detection and identification simply because new sequences can be added quickly and just as quickly be placed into operation, and the new microbes are detected in samples. This capability of MSP/ABOID is also useful because since it is software it can re-analyze old files and determine if any of the newly added microbe sequences lead to a new detections and identification.

3 *Apis*

Apis is a genus of the family Apidae – they are the honeybees and are classified as shown in Figure 3.1.

Since this book is on microbe diversity found on honeybees, it is appropriate and since all the samples analyzed are honeybees that the MSP/ABOid analysis should be able to detect the unique peptides associated with them. How much easier it is to associate microbes with the honeybee when you detect both the honeybee and microbe peptides at the same time and in the same sample.

Although the queen often lives for several years (three to four years), the general life expectancy for the workers is only a few weeks. Thinking on this for a moment, it can be seen that the microbes that the workers collect are current and do not reflect a historic collection of microbes representing years of effort on the part of one bee. The concentrations of microbes can be expected to represent the current bioflora for the area in which the honeybees work.

When we talk about the microbe diversity in honeybees in the United States, we are really talking about microbes collected by one species – *Apis mellifera*. This was confirmed by analysis.

Let us consider the relationship of *Apis mellifera* and other species. It is of interest to consider other species of honeybees only to show that they can also be detected. Although it is beyond the scope of this effort, it is certainly of interest to know what might be discovered by examining all the honeybee species worldwide in their native inhabitant. The detection of other species of honeybee might also be useful to determine their current range and to observe a movement of one species into other areas of the world. This could also be useful for monitoring the movement of their main product – honey. Likewise, perhaps the world-wide biome could be evaluated by analyzing the collections of microbes by honeybees. Over time these collections of data could be used to evaluate environmental changes that occur naturally over a year, or seasons, and be a measure of a baseline to monitor changes caused by natural and manmade influences such as weather and urbanization.

3.1 THE GENUS *APIS*

As mentioned, within the genus *Apis* there are seven to twelve species of which only four are sufficiently sequenced to serve in a current data group. Additional species can be added as they are sequenced. These four will be discussed as they all contribute to a discussion on detecting the honeybee genome.

Domain: Eukaryota

Kingdom: Metazoa

Phylum: Arthropoda

Subphylum: Uniramia

Class: Insecta

Order: Hymenoptera

Family: Apidae

Genus: *Apis*

Species: *Apis mellifera*

FIGURE 3.1 Classification of *Apis mellifera.*

3.1.1 *Apis Cerana*

Apis cerana, is known as the Asiatic honeybee or the eastern honeybee. It is native to South, Southeastern and East Asia. There are eight subspecies.

The biology of *Apis cerana* F., the Asian honey bee, is far less known than that of its sister species *Apis mellifera* L. The arrival of *A. cerana* in North Queensland has prompted the need to better understand the ecology and biology of this species in an invasive context (Guez, Subias, & Griffin, 2017).

3.1.2 *Apis dorsata*

Apis dorsata, or the Asiatic giant honeybee, is found throughout South and Southeast Asia where they are seasonally abundant throughout the upland and lowland rainforests in Southeast Asia. The colonies of *Apis dorsata* are found to nest in aggregates on tall bee trees (tree emergent) in the open, as well as, nesting singly in concealed locations when nesting low, especially in the submerged forest of *Melaleuca cajuputi* as in the vast area of the Melaleuca forest along the coastal areas of Terengganu (Saberioon, Mardan, Nordin, Alias, & Gholizadeh, 2010).

3.1.3 *Apis florea*

Apis florea, is known as the dwarf honeybee located in South and Southeast Asia.

Identification of pollen grains from Thai honeybees, *Apis cerana, A. dorsata* and *A. florea* in Nan province, northern Thailand, was used to determine foraging success (Suwannapong, Maksong, Yemor, Junsuri, & Benbow, 2013).

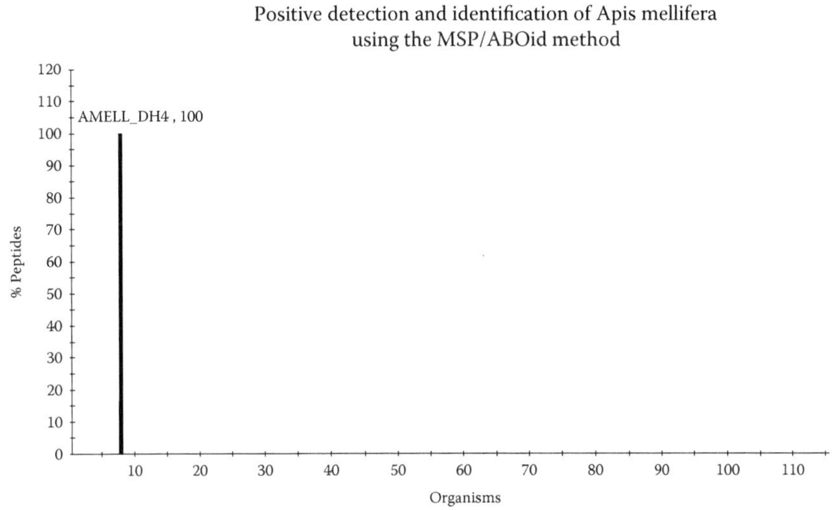

FIGURE 3.2 Detection and identification of *Apis mellifera* using the MSP/ABOid method.

Along the Nile river, *A. florea* has a significantly higher population density than the wild, native *A. mellifera.* Although *A. florea* is considered the invasive species, no indication of competitive displacement has been observed, and it appears that it coexists with the native *A. mellifera* (El-Niweiri, Moritz, & Lattorff, 2019).

3.1.4 *Apis mellifera*

Apis mellifera is the Western honeybee native to Europe, western Asia and Africa. They are central-place foragers and are known to forage nearly a mile from their hives. This long forage range is helpful when looking at how they travel, how they collect nectar and pollen, but is particularly important when studying the microbes of a particular area using the honeybee as a collector.

3.2 DETECTING *APIS*

Figure 3.2 contains the MSP/ABOid detection and identification of *Apis mellifera* and Figure 3.3 is the chart with the identification and detection of *Apis cerana*, *Apis dorsata* and *Apis florea*. For the purpose of this demonstration, two different samples are used in Figure 3.3: a sample from the United States and one outside the United States. It is confirmation that *Apis mellifera* is present within the United States and all four types of honeybees are present outside the United States.

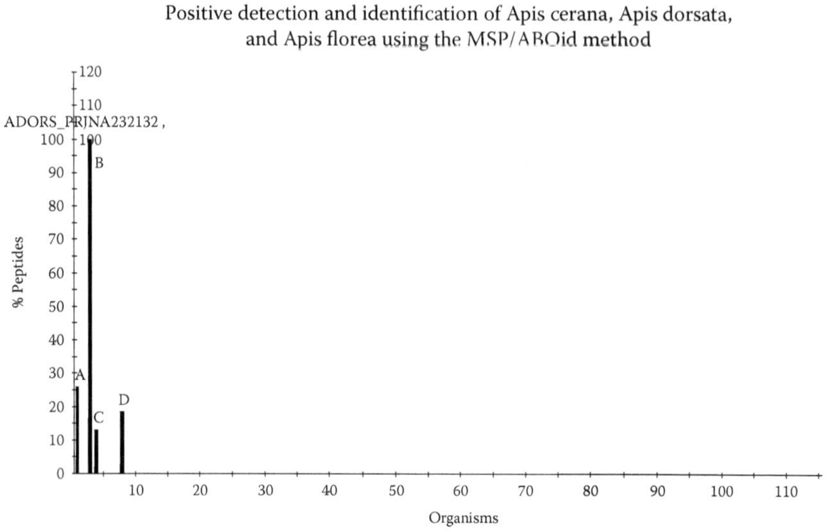

FIGURE 3.3 Detection and identification of *Apis cerana (A)*, *Apis dorsata (B)* and *Apis florea (C) and Apis mellifera (D)* using the MSP/ABOid method.

3.3 DISCUSSION

All samples in this collection contain honeybees. The samples represent more than 1,000 samples from around the United States, and the expectation was that we are collecting and analyzing *Apis mellifera*. It is useful to be able to detect and identify other species of *Apis*. Therefore, a foreign sample was collected and analyzed to test the ability to detect other species of *Apis*. The results are shown in Figure 3.2 and Figure 3.3 which demonstrated the detection of four *Apis* species: *Apis cerana*, *Apis dorsata*, *Apis florea* and *Apis mellifera*. As a result of this capability, all our samples verified the detection of *Apis mellifera*.

4 Bee Gut

4.1 BEE GUT MICROBES

Bee gut microbes are presented exclusively in this chapter because of their importance to the honeybee. These microbes work together within a complex metabolic relationship that allows the honeybee to digest a pollen-rich diet (Kešnerová et al., 2017). For example, nectar which is usually in a liquid form and mostly contains monosaccharide sugars is easily absorbed, while pollen contains complex polysaccharides: branching pectin and hemicellulose. To make this functional and highly efficient, five groups of bacteria take part in the digestion process. Three species of bacteria specialize in digesting simple sugars (*Snodgrassella* and two species of *Lactobacillus*), and two species specialize in digesting complex sugars (*Gilliamella* and *Bifidobacterium*). Each may contribute to the success of each other by producing products that are beneficial to all, such as the production of hydrogen peroxide by *Gilliamella*. Likewise, the digestion of pectin by *Gilliamella* and the digestion of hemicellulose by *Bifidobacterium* contribute to the digestion process. If one microbe cannot accomplish a function, it can use enzymes from their neighbors to accomplish digestion and probably other functions while coexisting in the bee gut (Zheng et al., 2019).

Each of the major bee gut microbe is discussed in the following sections with analysis of their function and relationship to each other as demonstrated by the figures that illustrate the number of unique peptides for each microbe. Without knowing any further information, it is interesting to see that the *Bifidobacterium*, *Gilliamella* and *Lactobacillus* species are all within about the same number of unique peptides detected. *Serratia marcescens*, on the other hand, has almost twice as many peptides.

4.2 NATIONAL AVERAGE FOR BEE GUT MICROBES

The national average for bee gut microbes (Figure 4.1) illustrates three things: first, the average of unique peptides ranges between 10 and 20; second, *Serratia marcescens* ranges between 20 and 40; third, *Lactobacillus johnsonii* is very low.

4.2.1 *BARTONELLA APIS*

Bartonella apis is a gram-negative, rod-shaped bacterium found only in the gut of the honeybee *Apis mellifera* and is the only genus in the family *Bartonellaceae* (Kešnerová, Moritz, & Engel, 2016, *Bartonella apis* sp. nov., a

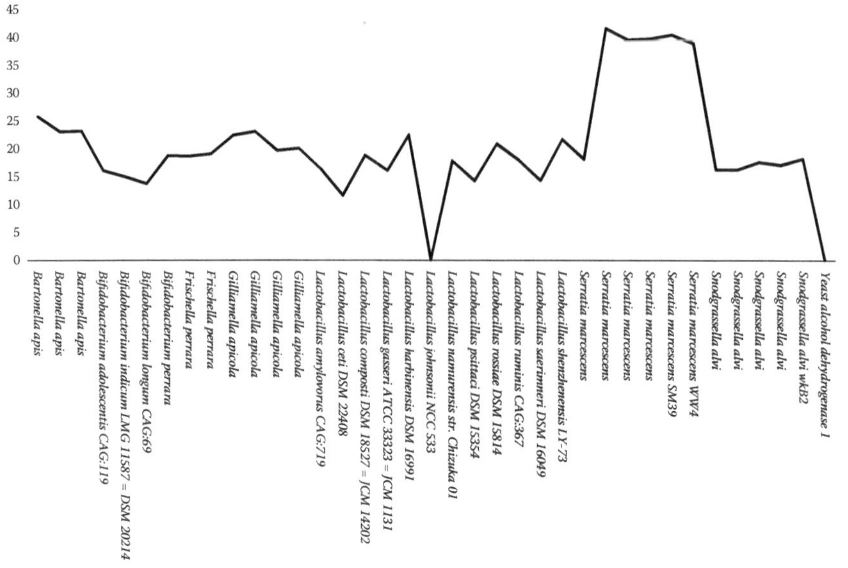

FIGURE 4.1 National average for bee gut microbes.

honeybee gut symbiont of the class *Alphaproteobacteria*). Figure 4.2 illustrates the number of unique peptides for *Bartonella apis* (three species) as found in the honeybees. The national average is 25.8 unique peptides. The California average is 18.1, and the average for Florida is 12.2), Idaho 25–30, Iowa 37–41.8 and Montana 47–49.

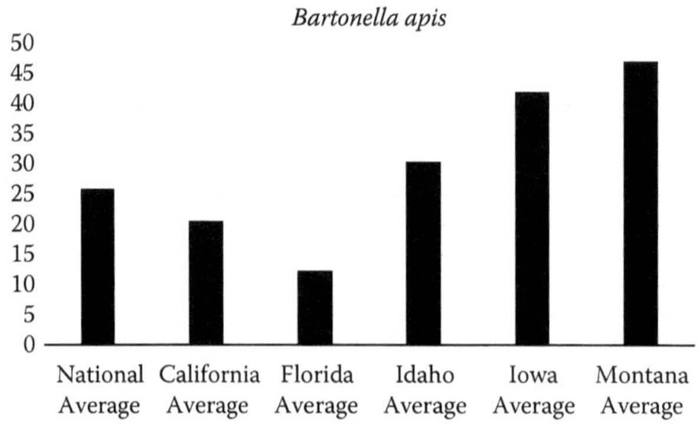

FIGURE 4.2 National and regional unique peptide averages for *Bartonella apis* using the MSP/ABOid method.

4.2.2 *Bifidobacterium indicum*

Bifidobacterium sp. is a gram-positive, rod-shaped bacterium. It is associated with the gut of most social insects – honeybees, wasps, cockroaches and bumblebees. In honeybees, the counts of *Bifidobacteria* can be 2%–8% of the total bacteria (Scardovi & Trovatelli, 1969).

Figure 4.3 shows the national and regional averages for three species of *Bifidobacterium*: *B. adolescentis*, *B. indicum* and *B. longum*. Examining further, Figure 4.1 illustrates the national average for the unique peptides found in honeybees as well as the averages for five regions. The national average ranges from 13.8 to 16.2 for the three species with *B. adolescentis* having 16.2 and *B. longum* having 13.8 unique peptides. Florida has the lowest overall number of unique peptides for all three species ranging from 7.1 to 7.3, and Montana has the highest ranging from 29.8 unique peptides for *B. adolescentis*, followed by *B. indicum* with 26.8 and *B. longum* with 26.5.

4.2.3 *Gilliamella apicola*

Gilliamella apicola is a gram-negative, rod-shaped bacterium, microaerophile gut symbiont, found in the honeybee gut microbiota, wherein it utilizes several sugars that are harmful to the bee host.

The national average ranges from 19.7 to 23.1 unique peptides for the four strains of *Gilliamella apicola* (Figure 4.4). The Florida region has the lowest average ranging from 9 to 11 unique peptides, while Montana has the highest

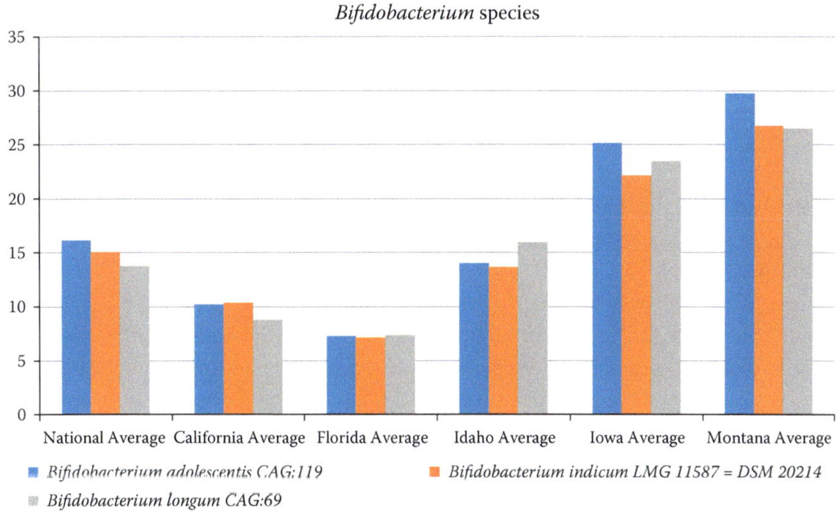

FIGURE 4.3 National and regional unique peptide averages for *Bifidobacter* using the MSP/ABOid method.

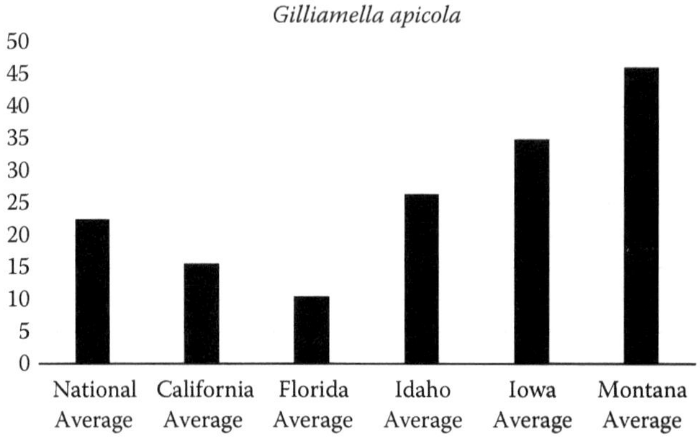

FIGURE 4.4 National and regional unique peptide averages for four species of *Gilliamella* using the MSP/ABOid method.

ranging from 35.3 to 50.5. It is interesting to note that two species in the Montana average dominate.

4.2.4 *Lactobacillus*

Lactobacillus is a large genus of gram-positive, facultative anaerobic or microaerophilic, rod-shaped, non-spore-forming bacteria. *Lactobacillus amylovorus* is found in the gut of honeybees where they convert sugars into lactic acid as a by-product of glucose metabolism.

Lactobacillus acidophilus was recognized as a heterogeneous group by DNA analysis, eventually forming six distinct species composed of *L. acidophilus*, *L. amylovorus*, *Lactobacillus crispatus*, *Lactobacillus gallinarum*, *Lactobacillus gasseri*, and *Lactobacillus johnsonii* (Selle, Klaenhammer, & Russell, 2014).

L. johnsonii frequents the probiotic discussion elsewhere, but since it is an important species it is included in the bee gut data group. It is interesting in that *L. johnsonii* was not found in the bee gut results as seen in Figure 4.1 and Figure 4.5.

Lactobacillus bacteria are widely distributed in animal feeds, silage, manure and milk and milk products. Various species of *Lactobacillus* are used commercially during the production of sour milk, cheeses and yogurt, and they have an important role in the manufacture of fermented vegetables (pickles and sauerkraut), beverages (wine and juices), sourdough breads and some sausages.

Examination of Figure 4.5 reveals a different value for the unique peptides among 12 different species of *Lactobacillus*. In the national average, eight species predominate: *amylovorus*, *composti*, *harbinensis*, *namurensis*, *rossiae*, *ruminis* and *shenzhenensis*. *Lactobacillus ceti*, *gasseri*, *psittaci* and *saerimner* are less obvious. This pattern follows for California, Iowa and Montana. *L. gasseri* and *psittaci* gain

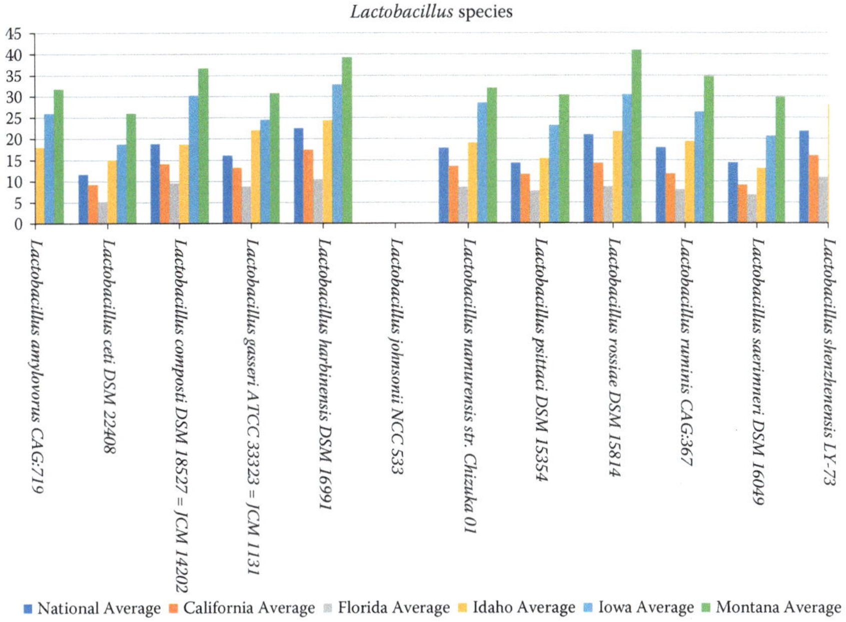

FIGURE 4.5 National and regional unique peptide averages for 12 species of *Lactobacillus* using the MSP/ABOid method.

in relation to the obvious in the Florida averages. *L. gasseri* is more abundant in the Idaho averages. Recall the discussion of the interaction and interrelationships between gut microbes; perhaps we are seeing different types of activity, such as a nectar diet or a pollen diet. Observing the levels of these bacteria in a honeybee sample could be helpful in understanding what type of activity is taking place in the honeybee at the given time. It would be interesting to follow this activity over a period of time or even over several seasons to determine the relationship of the *Lactobacillus* bacteria and the external factors, such as climate, environment, pollen, nectar sources, and so on. It could be sensitive enough to evaluate the impact of different food sources, including natural and artificial.

Lactobacillus amylovorus. The national average number of unique peptides for *L. amylovorus* is 16.5. This bacterium is present in all the regional areas, with the lowest being in Florida with an average of 8.3 and the highest in Montana with an average of 31.8.

Lactobacillus gasseri. The national average number of unique peptides for *L. gasseri* is 16.1. This bacterium is present in all the regional areas, with the lowest being in Florida with an average of 8.8 and the highest in Montana with an average of 30.8. The Idaho average for *L. gasseri* had the largest relative change when compared with the other *Lactobacillus* species showing an average number of unique peptides of 22.0 when compared with *L. composti* with an average of 18.7.

In the other regions *L. gasseri* held a relative place, with the highest average of 30.8 in the Montana region.

As mentioned, this bacterium is one of the six distinct species derived from the *L. acidophilus* heterogeneous group and is found in the gut of honeybees where it is likely to convert sugars into lactic acid as a by-product of glucose metabolism. *L. gasseri* is an anaerobe known to produce hydrogen peroxide and is important in that it can ferment proteins and degrade oxalate (Arihara, 1998). These are useful properties when all the microbes are considered and how they work together in the bee gut digestion of pollen, nectar and other collected items.

Lactobacillus harbinensis. The national average number of unique peptides for *L. harbinensis* is 22.5, which is the highest among all the *Lactobacillus*. This bacterium is present in all the regional areas with the lowest being in Florida with an average of 10.5 and the highest in Montana with an average of 39.3. *L. harbinensis* has greater number of unique peptides for a species among the national and California averages. It becomes second behind *L. shenzhenensis* in the other regions including Florida.

Lactobacillus shenzhenensis LY-73. The national average number of unique peptides for *L. shenzhenensis* is 21.7, which is second among all the *Lactobacillus*. The largest increase in unique peptides for *L. shenzhenensis* occurs in the Idaho (28 unique peptides), Iowa (35.2 unique peptides) and Montana (42.8 unique peptides) regions.

4.2.5 *SERRATIA MARCESCENS*

Serratia marcescens is a motile, facultative anaerobe capable of growing in temperatures ranging from 5 to 40°C and in pH levels ranging from 5 to 9. There are numerous strains.

Serratia marcescens is an opportunistic pathogen of many plants and animals and is present in the bee gut of honeybees. Three *S. marcescens* strains have been isolated from the guts of honey bees and have been characterized. These results tend to confirm these earlier surveys. Surveys taken from samples from four locations in the United States indicate the presence of *S. marcescens* in the guts of over 60% of worker bees (Kasie Raymann, 2018). *S. marcescens* is one of the most prevalent microbes found in the honeybee samples as seen in the national average (Figure 4.1). Recall that these results represent *S. marcescens* detected and identified from honeybees and not just the honeybee gut. The honeybee gut microbes are included in the whole sample, and *S. marcescens* is known to be found elsewhere in the environment and indeed may be associated with many other things other than the bee gut.

Figure 4.6 illustrates six strains of *S. marcescens* in relation to each other. The national average has five strains that range from 39.0 to 41.7 and one strain with only 18.1 unique peptides. This pattern is reflected through the five regions, with Florida having the highest numbers of unique peptides ranging from 18.9 to 21.2

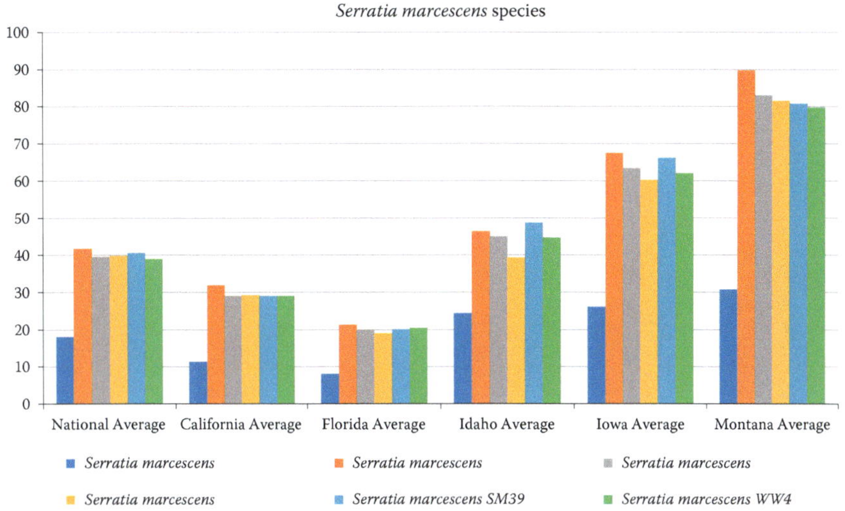

FIGURE 4.6 National and regional unique peptide averages for six strains of *Serratia marcescens* using the MSP/ABOid method.

and the lowest of 8.0. Montana has the highest numbers with a range of 79.8–89.8 for the five highest and 30.8 for the lowest.

Since *S. marcescens* is so prevalent, it would be useful to monitor the levels of this bacteria in honeybee or other samples taken in proximity to the honeybees. This information may indicate the current state of the environment, with higher numbers of *S. marcescens* indicating a potential issue and lower numbers being associated with good conditions. The sensitivity of microbes to changes in the environment may prove useful in assessing the relative health of the honeybee population.

4.2.6 *Snodgrassella alvi*

Snodgrassella alvi is a gram-negative, rod-shaped bacterium that forms smooth, white, round colonies on agar. *S. alvi* is found in the gut bacteria of the Western honeybee (*Apis mellifera*), where it forms a biofilm in the ileum (Kwong, 2012; Moran, 2012). The national Average for five strains of *Snodgrassella alvi* ranges from 16.3 to 18.2 unique peptides (Figure 4.7). The Florida region has the lowest average from 8.1 to 8.6, and Montana region has the highest numbers and the greatest range among the strains at 29.5–35.

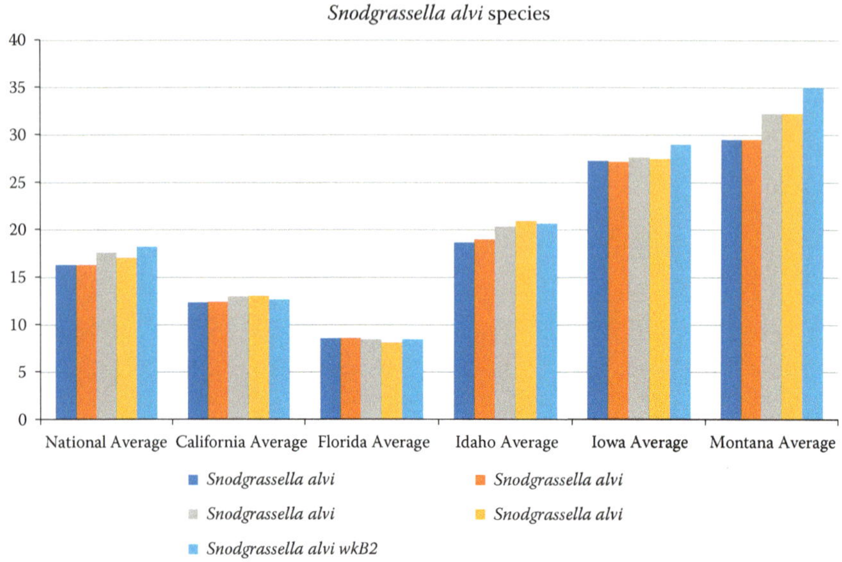

FIGURE 4.7 National and regional unique peptide averages for five strains of *Snodgrassella alvi* using the MSP/ABOid method.

4.3 REGIONAL AVERAGES

Each region, California, Florida, Idaho, Iowa and Montana, is graphed and compared with the national average.

4.3.1 California Regional Average

Comparison of the national average for bee gut microbes and the average for California is shown in Figure 4.8. The California average is lower, but follows the national average for all the bee gut microbes except *Gilliamella* which is even lower. In the national average, *Gilliamella* ranges from 12.8 to 15.5 for the four strains. The California average ranges from 20.1 to 23.1 for the same strains.

4.3.2 Florida Regional Average

The Florida average number of unique peptides for the bee gut microbes is lower for all microbes than the national average (Figure 4.9). The largest difference is the *S. marcescens* strains, which are about 20 unique peptides lower than the national average.

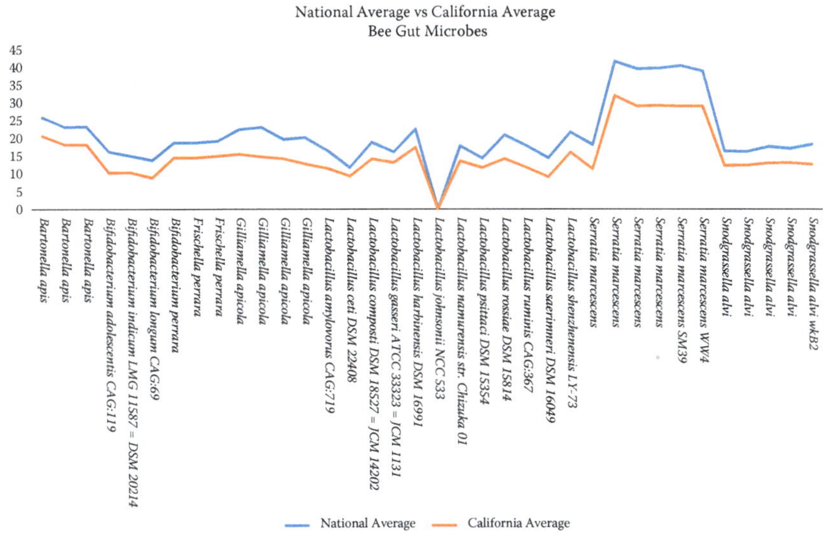

FIGURE 4.8 National and bee gut microbe unique peptide averages for California region using the MSP/ABOid method.

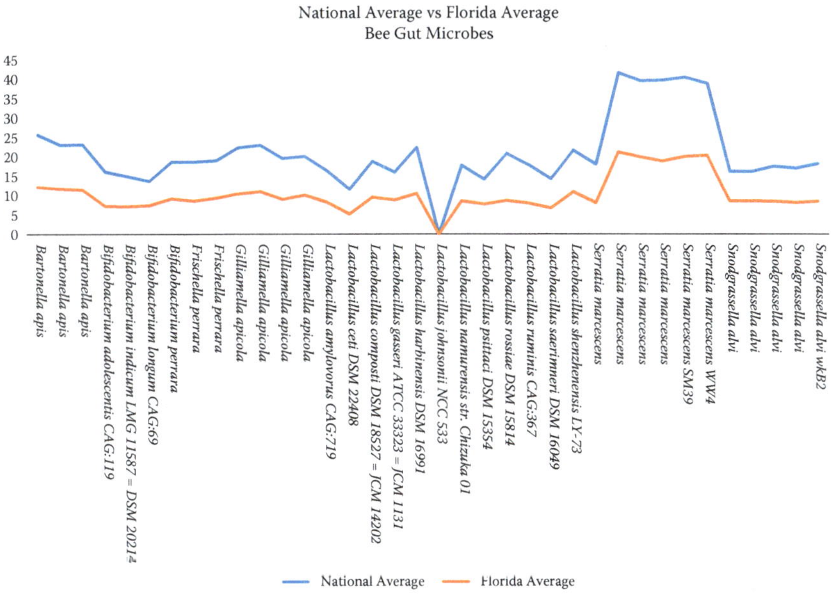

FIGURE 4.9 National and bee gut microbe unique peptide averages for Florida region using the MSP/ABOid method.

4.3.3 Idaho Regional Average

The Idaho average number of unique peptides for the bee gut microbes is nearly the same as the national average (Figure 4.10).

4.3.4 Iowa Regional Average

The Iowa average number of unique peptides for the bee gut microbes is greater than the national average. *Serratia sp.* are all higher than the national average in some cases by more than 20 peptides (Figure 4.11).

4.3.5 Montana Regional Average

The Montana average of unique peptides for many bee gut microbes is higher than the national average (Figure 4.12). Starting with *B. apis*, the national average is 23.2 and the Montana average is 49.0, a change of more than 25 unique peptides. The same holds for *G. apicola*, where the national average is 23.1 and the Montana average is 50.5, a difference of more than 27 unique peptides. The next large difference is the *Serratia sp.* where the national average ranges from 39 to 41.7 and the Montana average ranges from 79.8 to 89.8, a change of more than 40 unique peptides.

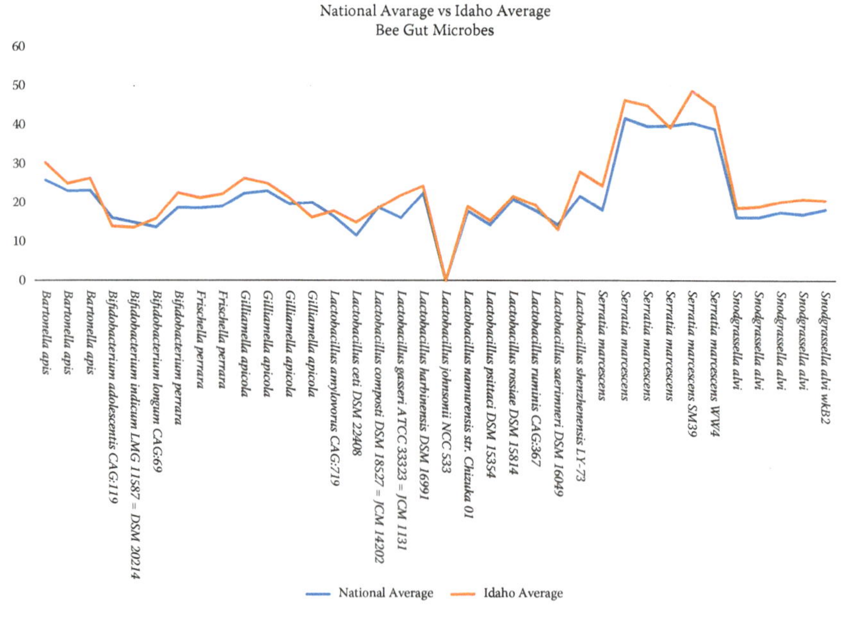

FIGURE 4.10 National and bee gut microbe unique peptide averages for Idaho region using the MSP/ABOid method.

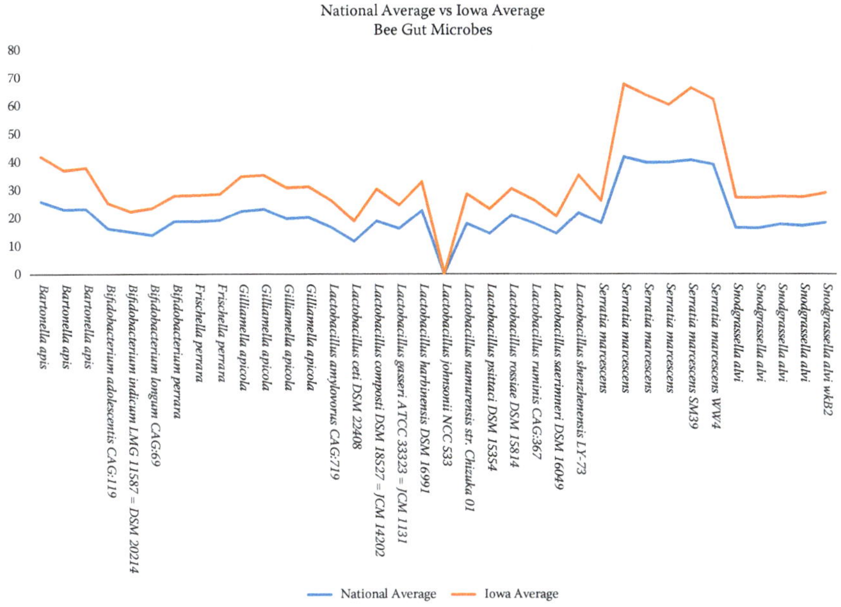

FIGURE 4.11 National and bee gut microbe unique peptide averages for Iowa region using the MSP/ABOid method.

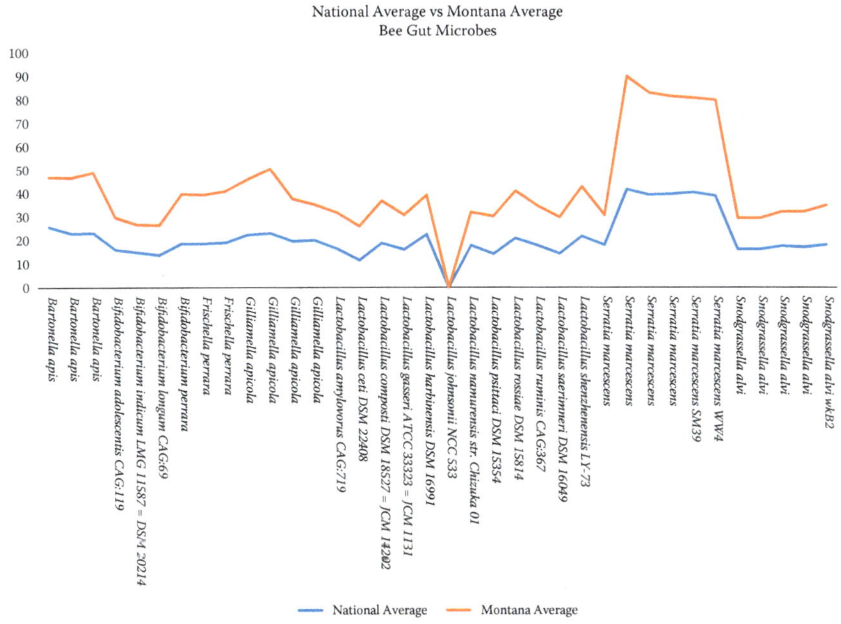

FIGURE 4.12 National and bee gut microbe unique peptide averages for Montana region using the MSP/ABOid method.

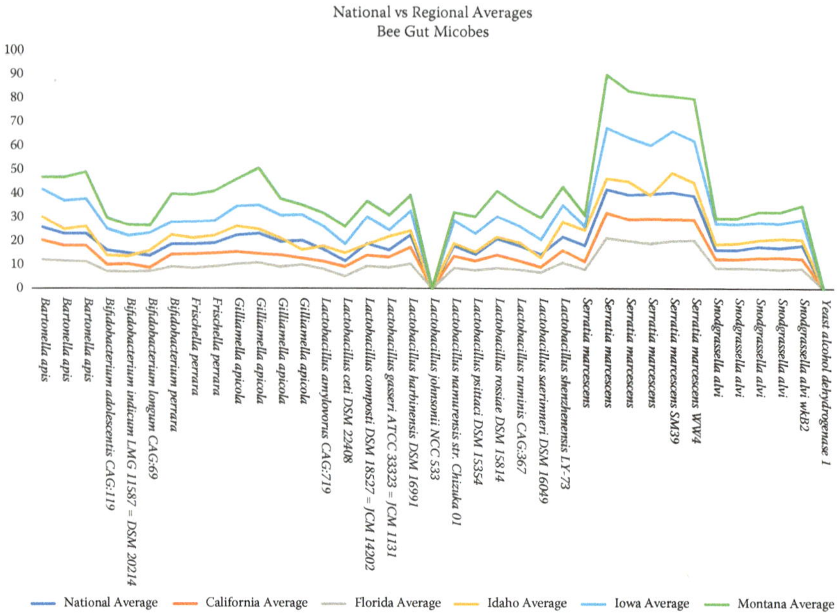

FIGURE 4.13 National and regional average summary of unique bee gut microbe peptides using the MSP/ABOid method.

4.3.6 National and Regional Averages for Bee Gut Microbes

The obvious change in the summary of all regions and the national average is seen in Figure 4.13, where the *Serratia sp.* has the greatest changes.

4.4 DISCUSSION

Six microbes and several strains of these microbes associated with the honeybee gut were examined in this chapter. More specifically, these were the microbes that were detected and identified from the honeybee samples. It is interesting to see the higher number of unique peptides detected for each species and the change in numbers between the different regions. The meaning of regional fluctuations and similar efforts is of interest but beyond the scope of this work.

Several useful capabilities result from this effort. Foremost, the ability to easily add new microbes as they are discovered and their sequences is made available. When changes to bee gut microbes are determined, their sequences can be downloaded from the National Center for Biotechnology Information and added to the bee gut data group. The same computer files that were used in this analysis can then be re-analyzed yielding new information. The results should prove interesting.

Likewise, a review of seasonal or yearly samples may prove of value in following these microbes and their association with honeybee activity in different environments and situations such as overwintering.

5 Coronavirus

In December 2019 a new severe acute respiratory syndrome (SARS) coronavirus was identified as infecting humans. Specifically, this particular severe acute respiratory syndrome coronavirus 2 was found to be a novel coronavirus called "SARS-CoV-2" (previously referred to as "2019-nCoV"), and it was determined to be a new strain that had not previously been identified in humans. The disease that is caused by SARS-CoV-2 is called "COVID-19". This is explained in terms that COVID-19 stands for corona (CO) virus (VI) disease (D) 19 (2019 – the year that the virus was detected).

COVID-19, like other viruses, has rapidly mutated to form different strains of the virus. For example, the strain in western Canada originated in Iran, as did the strain in New Zealand and Australia. The Iranian line originally came from China as did some infections in Australia. There are European pockets of the virus from China. COVID-19 arrived from Italy into South America and Mexico as did many of the UK infections. Some strains may have passed through the Netherlands and Belgium before arriving in the United Kingdom. Needless to say, COVID-19, like other viruses, tends to move around as people move around, mutates and produces new strains. During the beginning of the COVID-19 outbreak there were only a few (15) sequences for the strains of COVID-19, and this quickly grew to nearly 100 sequences in two months and continues to expand, not unlike what was seen with the H1N1 influenza outbreak in 2008 (Wick et al., 2009).

In this application the sequence for SARS-CoV-2 isolate Wuhan-Hu-1, complete genome – GenBank: MN908947.3 was downloaded from the NCBI and put into the SARS data group and used for the analysis of honeybee samples. It was quickly discovered that as many more strains were being sequenced, one download was not going to be sufficient, therefore over 100 sequences were downloaded and integrated into the SARS data group. The SARS data group was then further augmented to include all the complete sequences for other coronaviruses, renamed to be the coronavirus data group with over 250 coronaviruses. The honeybee files were reexamined.

5.1 CORONAVIRUSES

Aside from the 100 SARS or COVID-19 coronaviruses, there are many others of interest. Several BAT coronaviruses exist in nature, as well as in bovine, goose, duck, feline, rat, sparrow, swine and turkey. The sequences for these coronaviruses have all been downloaded, added to the coronavirus data group and used to analyze the honeybee samples. A national average was determined as well as averages for five regions (Figure 5.1).

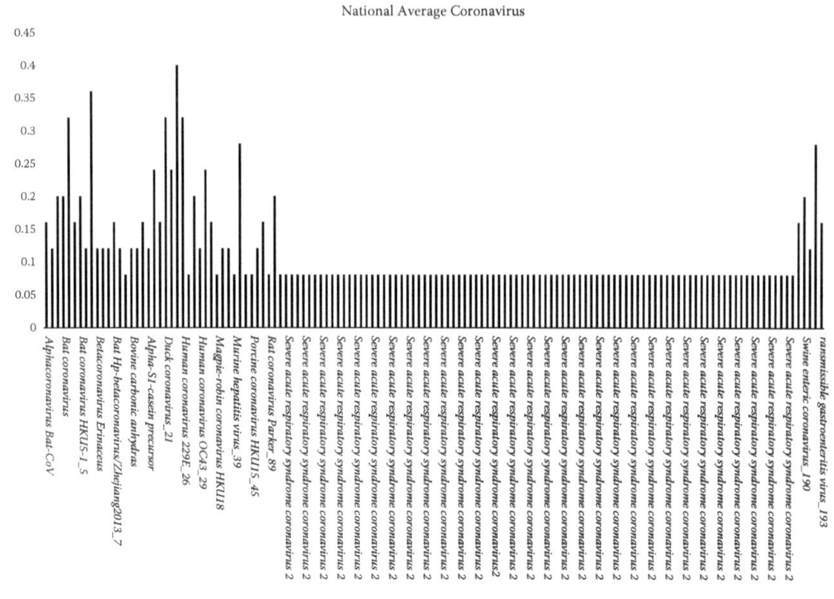

FIGURE 5.1 National unique peptide averages for coronavirus using the MSP/ABOid method.

5.2 NATIONAL AVERAGE FOR CORONAVIRUS

The initial result of analysis with the updated coronavirus data group was not very interesting with a large number of samples showing blanks. However, in April–June 2020, some honeybee samples were showing positive for COVID-19. The COVID-19 shows similar results, which may be due to lack of divergence among the strains; however, when looking at regional samples, and a smaller sample set, the results were different and averages were made for many coronaviruses. The regions, California, Florida, Idaho, Iowa and Montana all show results from the coronavirus data group. The reason for this is that although there were many blanks, there were enough samples that were positive to give an average. It should be emphasized that all these honeybee samples were analyzed using the same coronavirus data group. The regions are explored further.

5.3 CALIFORNIA REGIONAL AVERAGE

The California region illustrates three strains of bat coronavirus, one strain of bovine coronavirus, one strain of COVID-19, three strains of human coronavirus and one strain of rat coronavirus (Figure 5.2). Although the averages for the unique peptides are low, a detection and identification has been done for eight different coronaviruses, including one strain of COVID-19.

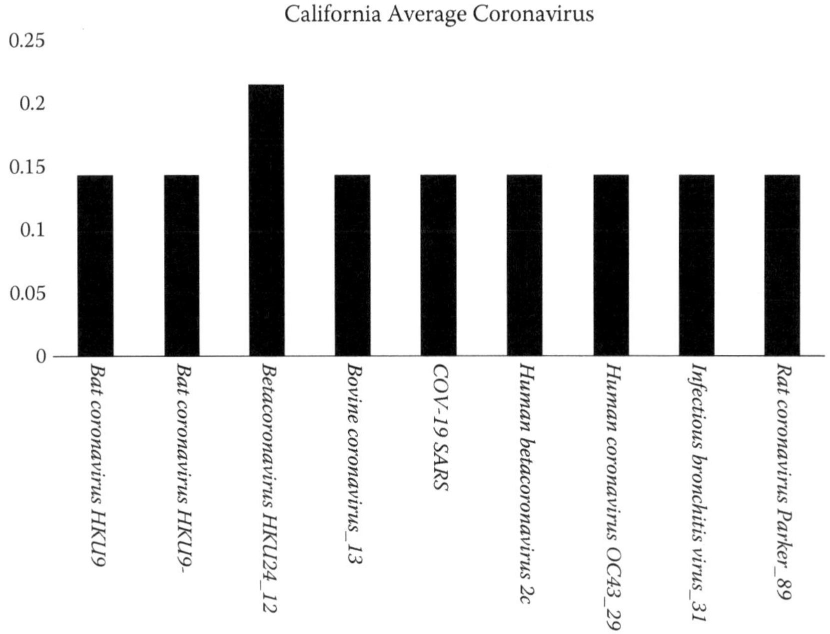

FIGURE 5.2 California region unique peptide averages for coronavirus using the MSP/ABOid method.

5.4 FLORIDA REGIONAL AVERAGE

The Florida region illustrates four strains of bat coronavirus, one strain of Canadian goose coronavirus, one strain of COVID-19, one strain of duck coronavirus, one strain of feline coronavirus, two strains of human coronavirus (one is infectious bronchitis virus), one strain of mink coronavirus and one strain of rat coronavirus (Figure 5.3). The averages for the number of unique peptides are low, but a detection and identification has been made for eight different coronaviruses, including one strain of COVID-19.

5.5 IDAHO REGIONAL AVERAGE

The Idaho region illustrates of one strain of bat coronavirus, one strain of human coronavirus, one strain of rabbit coronavirus, one strain of rodent coronavirus and one strain of turkey coronavirus (Figure 5.4). The averages for the number of unique peptides are low, but a detection and identification has been made for eight different coronaviruses and no strains of COVID-19.

5.6 IOWA REGIONAL AVERAGE

The Iowa region illustrates of five strains of bat coronavirus, one strain of Canadian goose, one strain of feline infectious peritonitis virus, one strain of

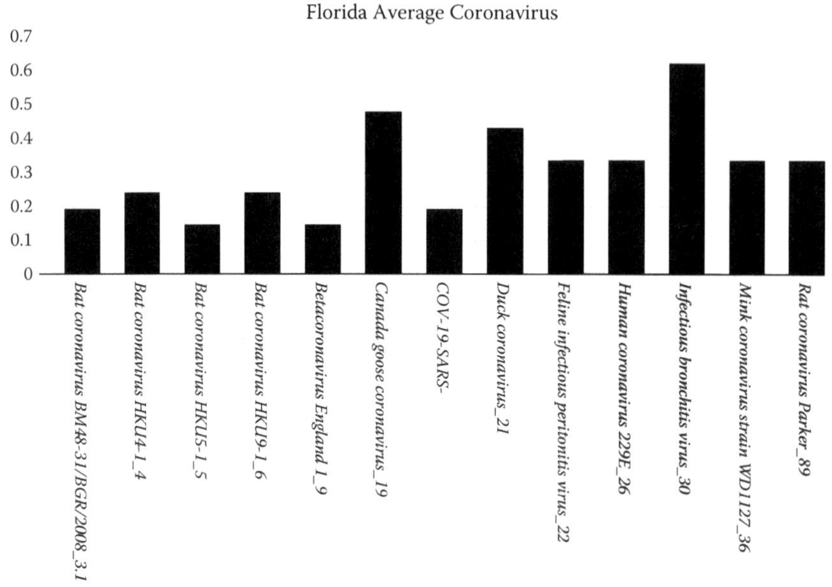

FIGURE 5.3 Florida region unique peptide averages for coronavirus using the MSP/ABOid method.

ferret coronavirus, one strain of human coronavirus, three strains of infectious bronchitis, one strain of mink coronavirus, one strain of munia coronavirus, one strain of night heron coronavirus, and two strains of rat coronavirus. There are 92 COVID-19 strains detected, with a couple strains showing a lower average; the similarity in unique peptides is most likely due to a lack of divergence among the strains. There is a strain of swine coronavirus, a strain of thrush coronavirus, a strain of turkey coronavirus and one strain of transmissible gastroenteritis virus (Figure 5.5). The averages for the number of unique peptides are low, but a detection and identification has been made for eight different coronaviruses, and 100 strains of COVID-19.

The Iowa region is different than the other regions. The COVID-19 strains all show a positive detection and identification. Some strains are of different averages. This region shows COVID-19 activity above the other regions. The other coronaviruses in this region look similar to those in the other regions with the Idaho region having the lowest number of different coronaviruses.

5.7 MONTANA REGIONAL AVERAGE

The Montana region illustrates five strains of bat coronavirus of different averages of unique peptides, one strain of bovine virus, one strain of Canadian goose, one strain of duck, one strain of Feline, four strains of

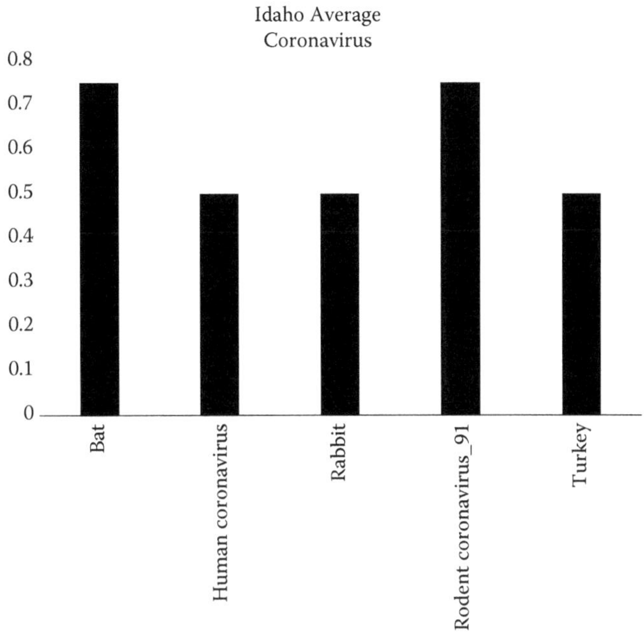

FIGURE 5.4 Idaho region unique peptide averages for coronavirus using the MSP/ABOid method.

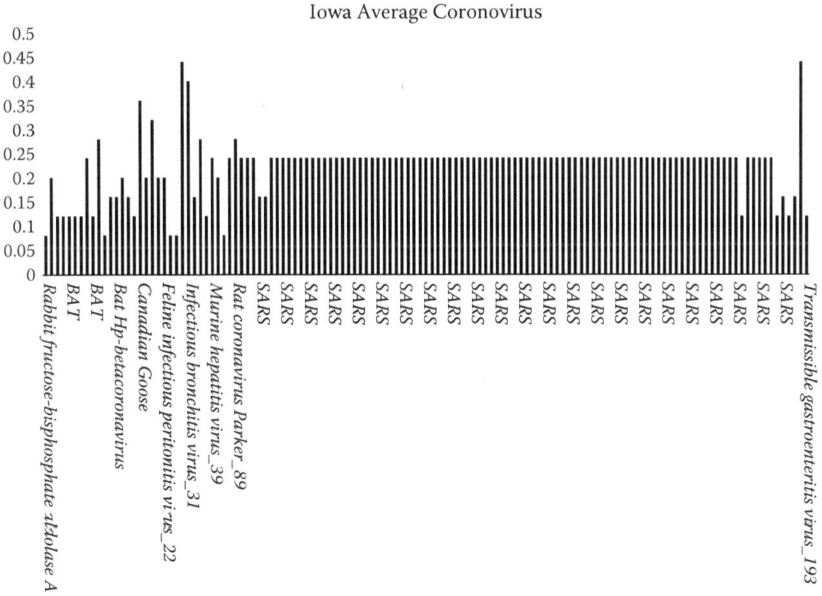

FIGURE 5.5 Iowa region unique peptide averages for coronavirus using the MSP/ABOid method.

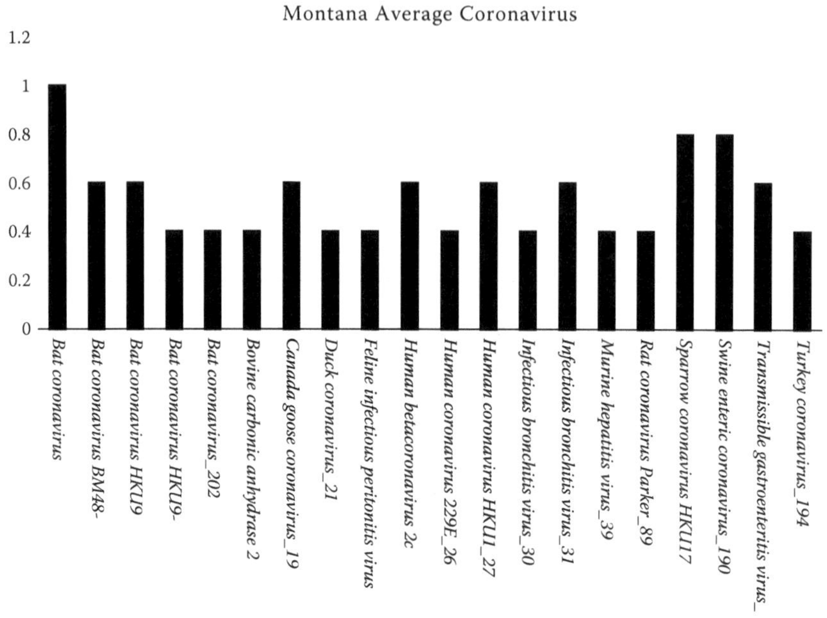

FIGURE 5.6 Montana region unique peptide averages for coronavirus using the MSP/ABOid method.

human coronavirus (one is Infectious bronchitis virus), one strain of murine hepatitis virus, one strain of rat coronavirus, one strain of sparrow coronavirus, one strain of swine coronavirus and one strain of turkey coronavirus (Figure 5.6). There is no detection of COVID-19. The averages for number of unique peptides are low, but a detection and identification has been made for 18 different coronaviruses strain and no strains of COVID-19.

5.8 VERIFYING COVID-19 DETECTION

Figure 5.7 illustrates the detection and identification of a lab-supplied COVID-19 sample. The main peak is the virus, the smaller peaks are other items not of interest. This is a clear result of the MSP/ABOid analysis using the standard protocols for all the honeybee samples.

Figure 5.8 is a honeybee sample showing a positive detection and identification of COVID-19. The main peak is the virus, the smaller peaks are other items not of interest. This is a clear result of the MSP/ABOid analysis using the standard protocols for all the honeybee samples.

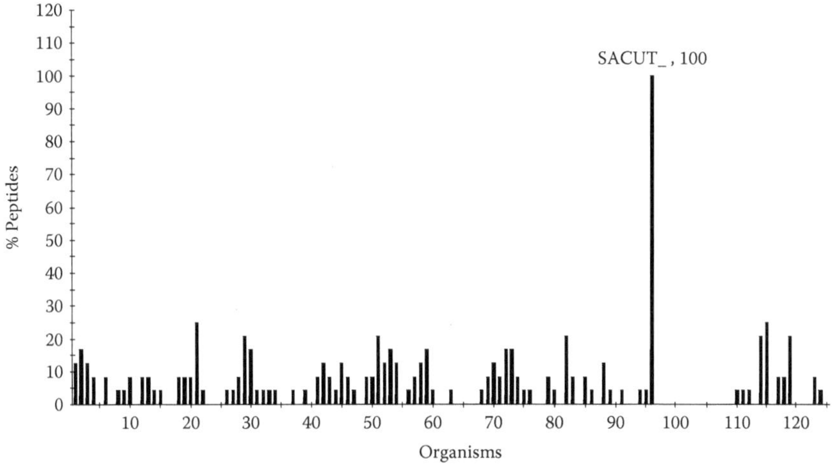

FIGURE 5.7 Positive detection and identification of a sample of COVID-19 from a supplier using the MSP/ABOid method.

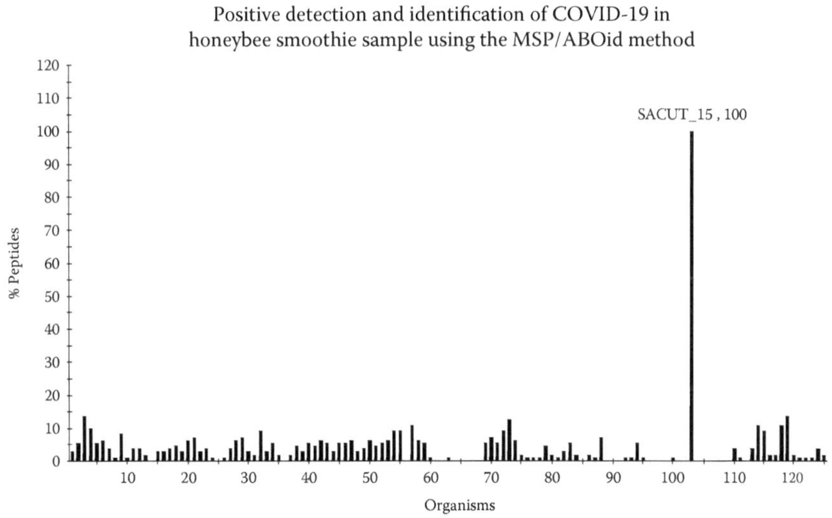

FIGURE 5.8 Honeybee smoothie sample with a detection and identification of COVID-19 using the MSP/ABOid method.

5.9 DISCUSSION

This chapter is of interest for many reasons other than demonstrating that coronaviruses can be detected from honeybees or rather that the honeybees collect many microbes from the environment and among those microbes are the coronaviruses, including COVID-19. The MSP/ABOid analysis using the standard protocols has shown to be successful in detecting a wide range of microbes, this just being one group. The distribution of the coronaviruses among the different regions illustrates that not all the regions have the same list and that they are different in type of virus and their frequency.

It was considered important to demonstrate that the MSP/ABOid analysis using the standard protocols could detect and identify the COVID-19. By taking a COVID-19 sample supplied by a laboratory and comparing it directly with a sample prepared from a honeybee sample demonstrates that the MSP/ABOid analysis is accurate.

The results from the different regions showed that a couple of the regions (Idaho and Montana) showed no detection of COVID-19, while showing a detection of a variety of other coronaviruses is telling. Likewise, the detection of one strain of COVID-19 in the California and Florida regions, as well as the detection of a variety of other coronavirus strains, is interesting. It appears it would be useful to monitor strains of human coronavirus and infectious bronchitis virus, in several of the regions as well as the COVID-19. There were not a sufficient number of samples to determine the averages for all the coronaviruses in other states, such as New York and other eastern areas or Texas and the southern states. It would seem appropriate to analyze honeybee samples from these areas to determine the averages. The spread of a virus such as COVID-19 could prove valuable in determining the spread and the relative averages for each state or pocket of interest, such as a town or area. This applies equally to monitoring the other coronaviruses.

6 Bacteria

Bacteria are physiologically diverse and ubiquitous in the world ecosystem. The bacteria data group used in this chapter contains several hundred sequenced bacteria. These have been acquired from the National Center for Biotechnology Information (NCBI) and incorporated into the bacteria data group which was used to identify the many types of bacteria reported. The honeybees collect a wide range of microbes on their forages and bacteria are among the microbes picked up and along for the ride. The bacteria picked up may be part of some other activity and the honeybee just happens to collect them while they are about their business collecting pollen and nectar. Likewise, bacteria may be found in the hive where the honeybee picks them up.

6.1 BACTERIA

Prokaryotes are divided into two kingdoms, namely bacteria and archaea. Bacteria have over 259,500 sequenced organisms divided into a plethora of groups, sub-groups, not to mention an extensive number of genera, species and strains. Bacteria have been around since the beginning of life on Earth and occur in nearly every niche in the living world.

For our purposes since we are reporting on the detection and identification of bacteria carried on the honeybee, it is sufficient to discuss a couple of simple terms that are used to separate or classify bacteria. One is the morphology, rod-shaped or round (cocci), the second is the gram stain. The gram stain was developed in 1884 by Hans Christian Gram. He developed a means to characterize bacteria based on their structural characteristic, namely the structure of their cell walls (Gram, 1884). The thick layers of peptidoglycan in the "gram-positive" cell walls stain them purple, while the thin "gram-negative" cell walls appear pink. With these two features, bacteria can be separated into these groups: gram-positive – rod shaped, gram-negative – rod shaped, gram-positive cocci and gram-negative cocci (Gram, 1884).

Most bacteria are gram-negative as a result of a cell wall that consists of peptidoglycan surrounded by a second lipid membrane containing lipopolysaccharides and lipoproteins. Gram-positive bacteria have thick cell walls containing many layers of peptidoglycan and teichoic acids. Firmicutes and Actinobacteria are Gram-positive (Hugenholtz, 2002).

Some bacteria have cell wall structures that are neither gram-positive nor gram-negative. These bacteria include those that have a thick peptidoglycan cell wall like a Gram-positive bacterium, but also a second outer layer of lipids (Alderwick, Harrison, Lloyd, & Birch, 2015). Considering the number of bacteria and the abundant manner in which they can diversify there are most likely other exceptions to this very simple classification.

Over 300 bacteria were detected and identified from honeybee samples from five regions that include California, Florida, Idaho, Iowa and Montana. From this collection of honeybees, an average was determined for each species of bacteria based on the number of unique peptides. A national average was calculated by averaging all the samples collected from all the regions.

6.2 NATIONAL AVERAGE FOR BACTERIA

The national average of unique peptides for 397 bacteria is shown in Figure 6.1. Looking at this interesting figure, there are 40 peaks that stand out from the rest. These species will be discussed next. The second feature of this figure is that the average unique peptides for each species is lower than other microbes, such as the fungi and viruses. This low average is also indicative of the results that show many samples had only a few bacteria peptides and others had an abundance. The samples with a few were in the majority. The diversity of the bacteria may be indicative of the nature of the distribution of bacteria in the areas where the honeybees operate. Looking over this long list it is tempting to examine all the different types of bacteria and attempt to group them according to some scheme and try and see if a pattern emerges for the areas where the honeybees were working. Soil, water and other factors could play a role or be a function in the equation that could explain or illuminate conditions in various areas. The honeybees bring these microbes back to their hive, and the colony then needs to sort them out by threat and deal with them. The large number of potential

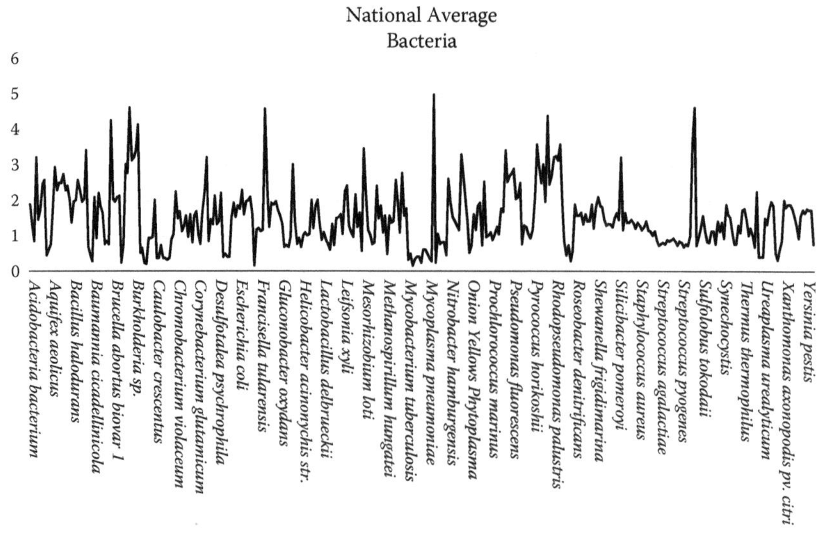

FIGURE 6.1 National unique peptide averages for Bacteria. Listing of all National bacteria and their average of unique peptides is given in Appendix A.

schemes that can be envisioned is beyond the scope of this book, but could supply ample information for a future effort. In the meanwhile, and particularly as new bacteria sequences are added to the bacteria data group, we will present what we have found and sorted through the national and regional averages which have a story to tell.

6.2.1 *Acidobacteria bacterium*

Acidobacteria is a phylum of diverse and ubiquitous gram-positive, aerobic bacteria (Euzeby, 1997). These bacteria are abundant in soil and can be expected to be part of any collection of microbes collected from soil. Because this is such a large phylum, members of the subdivisions are frequently found in many other areas including both high and low pH environments. Some members are able to use D-glucose and lactose as carbon sources with enzymes useful in the breakdown of sugars. This would indicate a possible connection with other microbes found in the honeybee collection that work together to breakdown sugars and the components of nectar and pollen.

The national average for this *Acidobacteria bacterium* is 1.9 unique peptides. The California average is 1.1 and the Florida average is 1.4. The Idaho average is 3.3 unique peptides and Iowa has 4.1 unique peptides (Figure 6.2). The most unique peptides for *A. bacterium* are in the Montana region with an average of 15.5. Examination of Figure 6.1, the national average for bacteria, illustrates that the Montana region has the highest average for all bacteria, and it is not surprising to see this ubiquitous gram-positive among the most numerous.

6.2.2 *Agrobacterium tumefaciens*

Agrobacterium tumefaciens is a gram-negative, rod-shaped bacterium of the phylum Proteobacteria. *A. tumefaciens* infects the plant roots causing Crown Gall Disease. It is closely related to nitrogen-fixing bacteria.

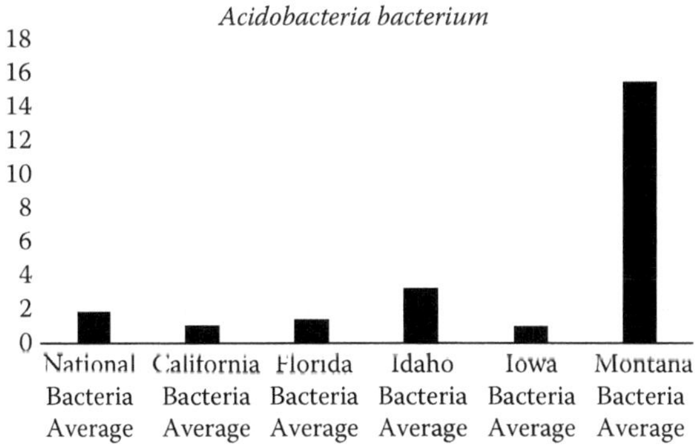

FIGURE 6.2 Detection and identification of *Acidobacteria bacterium* using the MSP/ABOid method.

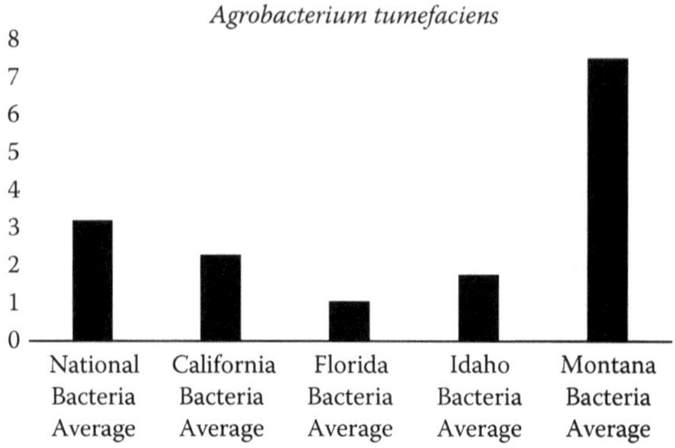

FIGURE 6.3 Detection and identification of *Agrobacterium tumefaciens* using the MSP/ABOid method.

Although closely associated with plants, *A. tumefaciens* has been shown to infect fungi and animals and most eukaryotic species by putting genes into a host. This feature could be of continuing interest in molecular biology technologies.

The national average for this *A. tumefaciens* is 3.2 unique peptides. The California average is 2.3 and the Florida average is 1.0. The Idaho average is 3.3 unique peptides and Iowa has 6.4 unique peptides (Figure 6.3). *A. tumefaciens* unique peptides are slightly higher in the Montana region with an average of 7.5. Examination of Figure 6.1, the national average for bacteria, illustrates that the Montana region has the highest average for all bacteria, and it is not surprising to see this gram-negative bacterium among the most numerous.

6.2.3 *Anabaena variabilis*

Anabaena variabilis is a bacteria strain of blue-green algae of the phylum Cyanobacteria. There are three distinct features of this strain: the ability of photosynthesis, nitrogen fixation and hydrogen production which allow *Anabaena variabilis* to live in different environments. It is found in aquatic environments and is believed as being a major producer of the oxygen in the atmosphere (Schaechter, 2006).

The national average for *A. variabilis* is 2.4 unique peptides. The California average is 1.9 and the Florida average is 0.9. The Idaho average is 2.8 unique peptides and Iowa has 5.3 unique peptides (Figure 6.4). *A. variabilis* unique peptides are slightly higher in the Montana region with an average of 9.5. Examination of Figure 6.1, the national average for bacteria, illustrates that the Montana region has the highest average for all bacteria, and it is not surprising to see this organism among the numerous.

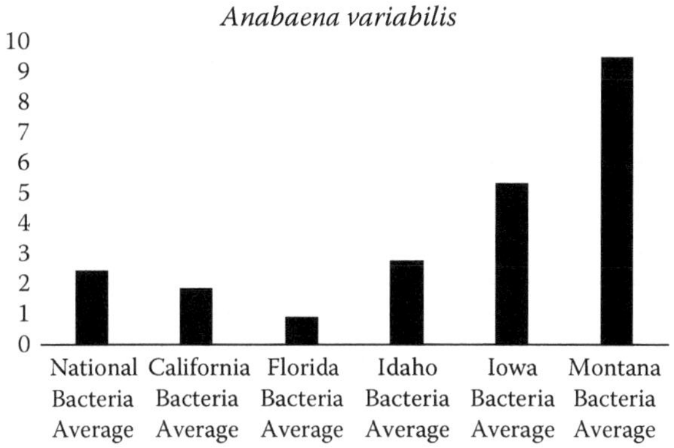

FIGURE 6.4 Detection and identification of *Anabaena variabilis* using the MSP/ABOid method.

6.2.4 *Azoarcus*

Azoarcus is a genus of nitrogen-fixing, rod-shaped bacteria of the phylum Proteobacteria (Reinhold-Hurek et al., 1993). Species of *Azoarcus* are motile by way of a single flagellum.

Azoarcus species are typically associated with plants as an aerobic and aromatic bacteria. They are also found in contaminated water and are common in the soil and grasses as well as some rice species. How this species interacts with the honeybee is not well known other than that the *Azoarcus sp.* are established in the environment and available for the honeybee to collect during foraging.

The national average for *Azoarcus* is 2.9 unique peptides. The California average is 1.9 and the Florida average is 0.8. The Idaho average is 2.0 unique peptides and Iowa has 3.9 unique peptides (Figure 6.5). *Azoarcus* unique

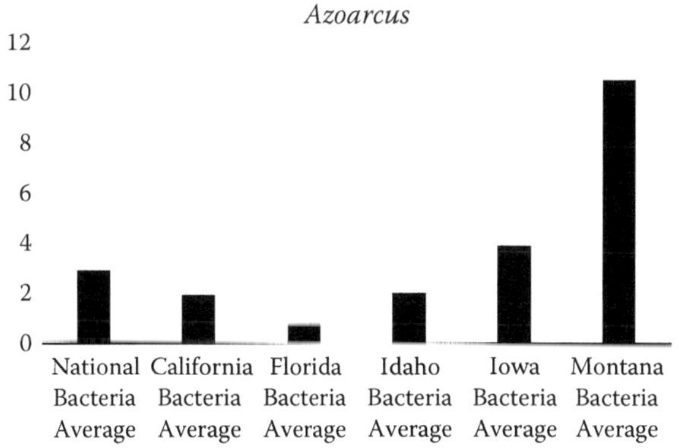

FIGURE 6.5 Detection and identification of *Azoarcus* using the MSP/ABOid method.

peptides are much higher in the Montana region with an average of 10.5. Examination of Figure 6.1, the national average for bacteria, illustrates that the Montana region has the highest average for all bacteria, and it is not surprising to see this organism among the numerous.

6.2.5 *Bacillus anthracis*

Bacillus anthracis is a gram-positive, rod-shaped, endospore-forming bacterium of the phylum Firmicutes. It lives in soils worldwide as a facultative anaerobe (can grow under aerobic or anaerobic conditions) and as an endospore-forming bacteria, can survive for long periods. There are 89 known strains of *Bacillus anthracis* with one being well known because it causes a disease in cattle known as "anthrax". *B. anthracis* belongs to the *B. cereus* group of strains.

The national average for *B. anthracis* is 2.5 unique peptides. The California average is 1.7 and the Florida average is 0.9. The Idaho average is 1.5 unique peptides and Iowa has 4.7 unique peptides (Figure 6.6). The unique peptides for *B.s anthracis* are much higher in the Montana region with an average of 7.0. Examination of Figure 6.1, the national average for bacteria, illustrates that the Montana region has the highest average for all bacteria, and it is not surprising to see this organism among the numerous.

6.2.6 *Bacillus cereus*

Bacillus cereus is a gram-positive, rod-shaped, facultatively anaerobic, motile, beta-hemolytic, spore forming bacterium of the phylum Firmicutes. It is commonly found in soil and food.

The *Bacillus cereus* group comprises six closely related species: *B. anthracis*, *B. cereus*, *B. cytotoxicus*, *B. mycoides*, *B. pseudomycoides*, and *B. thuringiensis*.

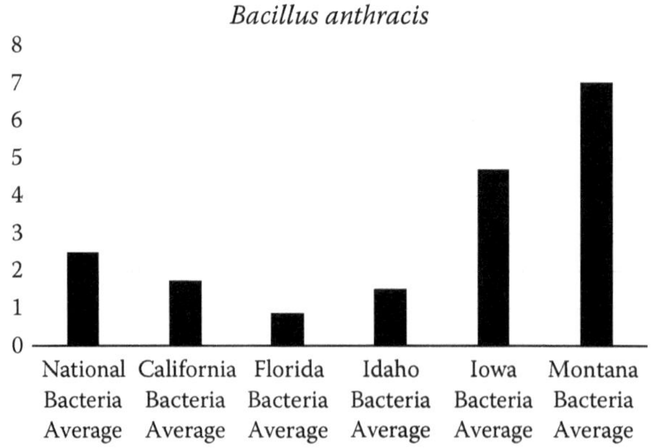

FIGURE 6.6 Detection and identification of *Bacillus anthracis* using the MSP/ABOid method.

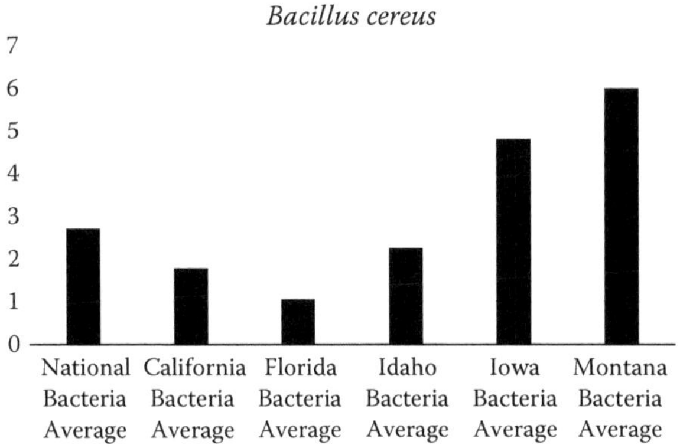

FIGURE 6.7 Detection and identification of *Bacillus cereus* using the MSP/ABOid method.

This group is widely distributed in nature and is commonly found in the soil as a saprophytic organism. *B. cereus* is also a contributor to the microflora of insects, and some plants (Vilain, 2006).

Some strains are used to protect alfalfa seeds, cucumbers, peanuts and tobacco seedlings and other plants as they naturally produce antibiotics that are useful in fighting a wide range of fungi and bacteria.

The national average for *B. cereus* is 2.7 unique peptides. The California average is 1.8 and the Florida average is 1.0. The Idaho average is 2.3 unique peptides and Iowa has 4.8 unique peptides (Figure 6.7). The unique peptides for *B. cereus* are much higher in the Montana region with an average of 6.0. Examination of Figure 6.1, the national average for bacteria, illustrates that the Montana region has the highest average for all bacteria, and it is not surprising to see this *Bacillus cereus* among the numerous microbes collected.

6.2.7 *Bacillus clausii*

Bacillus clausii is a gram-positive, motile, spore-forming, rod-shaped, alkaliphilic bacterium of the phylum Firmicutes. *B. clausii* is found in many alkaline environments, including soil and water.

B. clausii has been found to produce antimicrobial substances that are active against gram-positive bacteria including *Clostridium difficile*, *Enterococcus faecium* and *Staphylococcus aureus* (Urdaci, Bressollier, & Pinchuk, 2004). This activity can prove to be a benefit to a system of microbes as this capability could be helpful in keeping other bacteria under control while providing useful features to the system as a whole. In this regard, *Bacillus clausii* could be useful to the honeybee.

The national average for *B. clausii* is 2.0 unique peptides. The California average is 1.9 and the Florida average is 0.8. The Idaho average is 1.8 unique

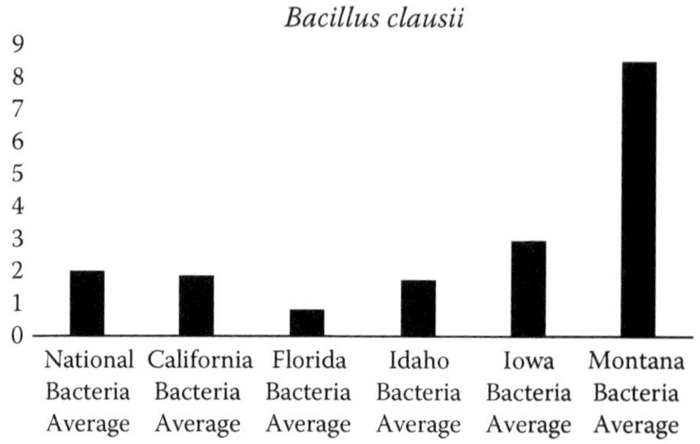

FIGURE 6.8 Detection and identification of *Bacillus clausii* using the MSP/ABOid method.

peptides and Iowa has 3.0 unique peptides (Figure 6.8). The unique peptides for *B. clausii* are much higher in the Montana region with an average of 8.5. Examination of Figure 6.1, the national average for bacteria, illustrates that the Montana region has the highest average for all bacteria, and it is not surprising to see the *B. clausii* among the numerous microbes.

6.2.8 *Bacillus licheniformis*

Bacillus licheniformis is a spore-forming soil organism of the phylum Firmicutes. It is a gram-positive, mesophilic bacterium that contributes to nutrient cycling and is known for nitrate reduction by producing a variety of extracellular enzymes. Combined with anaerobic growth and antifungal activity, this bacillus should be of interest in honeybee activities and the microflora that work together to promote honeybee health.

B. licheniformis is commonly found in soil and bird feathers. Birds that tend to stay on the ground more than the air (i.e., sparrows) and on the water (i.e., ducks) are common carriers of this bacterium. Since *B. licheniformis* is able to secrete serine protease, it has become of interest to commercial uses, such as laundry detergents and to control fruit diseases (Silimela & Korsten, 2007). *B. licheniformis* has showed an antagonistic effect to *Paenibacillus larvae* ATCC 9545, the causative agent of American foulbrood disease of honeybees (Alippi & Reynaldi, 2006). Combined with *B. cereus* (mv33), *B. subtilis* (m351) and others species of *Bacillus*, this antagonistic effect may be a helpful factor in the management of microbes in a honeybee hive.

The national average for *B. licheniformis* is 2.0 unique peptides. The California average is 1.4 and the Florida average is 1.0. The Idaho average is 1.5 unique peptides and Iowa has 3.8 unique peptides (Figure 6.9). The unique peptides for *B. licheniformis* are much higher in the Montana region with an average of 5.5. Examination of Figure 6.1, the national average for bacteria, illustrates that the

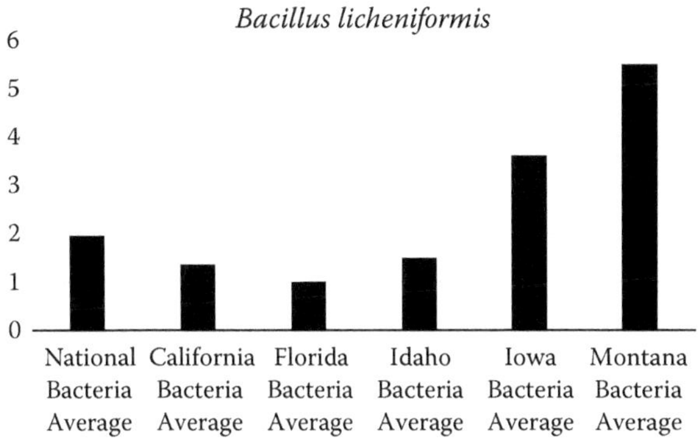

FIGURE 6.9 Detection and identification of *Bacillus licheniformis* using the MSP/ABOid method.

Montana region has the highest average for all bacteria, and it is not surprising to see *B. licheniformis* among the numerous, particularly since it can be isolated from practically anywhere because it produces highly resistant endospores.

6.2.9 *Bacillus subtilis*

Bacillus subtilis are rod-shaped, gram-positive, endospore-forming, motile, aerobic bacteria of the phylum Firmicutes. They are naturally found in soil and vegetation. Like other *Bacillus* species, *B. subtilis* has been used on plants as a fungicide, and some strains related to *B. subtilis* are capable of producing toxins for the control of insects. *B. thuringiensis* is a bacterium in the same genus that well known as an insecticide.

B. subtilis has been linked with higher elevations and can act as an identifier for both eco-adaptability and honey bee health (Sudhagar & Reddy, 2017).

The national average for *B. subtilis* is 2.0 unique peptides. The California average is 1.6 and the Florida average is 1.1. The Idaho average is 2.0 unique peptides and Iowa has 3.9 unique peptides (Figure 6.10). The unique peptides for *B. subtilis* are much higher in the Montana region with an average of 10.5. Examination of Figure 6.1, the national average for bacteria, illustrates that the Montana region has the highest average for all bacteria, and it is not surprising to see *B. subtilis* among the numerous, particularly since it can be isolated from practically anywhere because it produces highly resistant endospores.

6.2.10 *Bacillus thuringiensis*

Bacillus thuringiensis are rod-shaped, gram-positive, endospore-forming, aerobic bacteria of the phylum Firmicutes. They are naturally found in soil and vegetation. Like other *Bacillus* species, *B. thuringiensis* has been used on

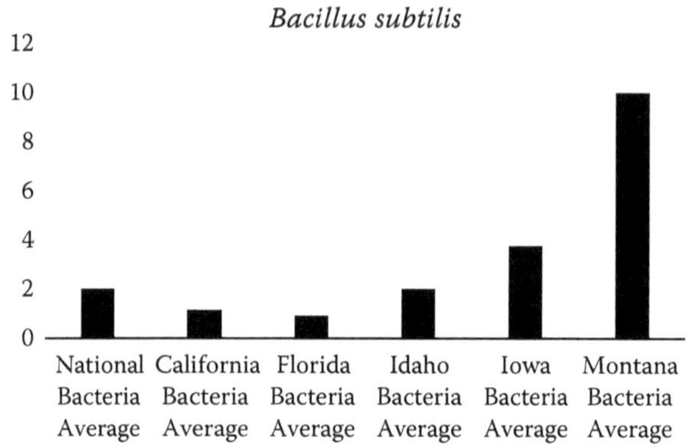

FIGURE 6.10 Detection and identification of *Bacillus subtilis* using the MSP/ABOid method.

plants as a fungicide, and some strains related to *B. thuringiensis* are capable of producing crystal proteins or "Cry" protein toxins for the control of insects.

The national average for *B. thuringiensis* is 2.6 unique peptides. The California average is 2.1 and the Florida average is 1.0. The Idaho average is 3.0 unique peptides and Iowa has 4.5 unique peptides (Figure 6.11). The unique peptides for *B. thuringiensis* are much higher in the Montana region with an average of 7.0. Examination of Figure 6.1, the national average for bacteria, illustrates that the Montana region has the highest average for all bacteria, and it is not surprising to see the *B. thuringiensis* among the numerous, particularly since it can be isolated from practically anywhere because it produces highly resistant endospores.

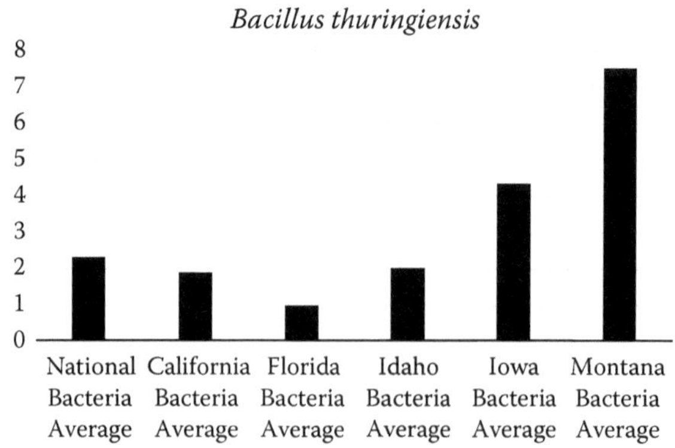

FIGURE 6.11 Detection and identification of *Bacillus thuringiensis* using the MSP/ABOid method.

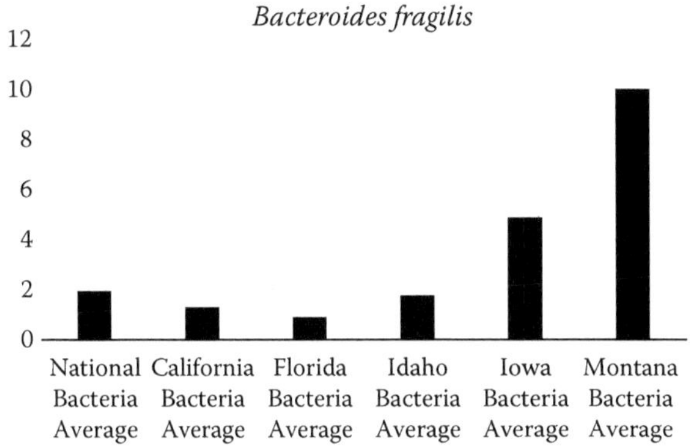

FIGURE 6.12 Detection and identification of *Bacteroides fragilis* using the MSP/ABOid method.

6.2.11 *Bacteroides fragilis*

Bacteroides fragilis is an obligately anaerobic, gram-negative, rod-shaped, non-spore-forming, non-motile bacterium of the phylum Bacteroidetes. It is part of the normal microbiota of the human colon.

Some enterotoxin-secreting strains of *B. fragilis* have been associated with diarrhea in piglets, calves, lambs, foals and humans, and these associations are considered to be a possible reservoir for the microbe and a likely source for the honeybee collections.

The national average for *B. fragilis* is 2.0 unique peptides. The California average is 1.3 and the Florida average is 0.9. The Idaho average is 1.8 unique peptides and Iowa has 4.9 unique peptides (Figure 6.12). The unique peptides for *B. fragilis* are much higher in the Montana region with an average of 10.0. Examination of Figure 6.1, the national average for bacteria, illustrates that the Montana region has the highest average for all bacteria.

6.2.12 *Bacteroides thetaiotaomicron*

Bacteroides thetaiotaomicron (formerly *Bacillus thetaiotaomicron*) is a gram-negative, obligate anaerobic bacterium of phylum Bacteroidetes. It is an opportunistic pathogen and second most common species isolated from the human gut flora, behind *B. fragilis* (Snydman et al., 2010).

The national average for *B. thetaiotaomicron* is 3.4 unique peptides. The California average is 2.5 and the Florida average is 1.5. The Idaho average is 2.0 unique peptides and Iowa has 8.2 unique peptides (Figure 6.13). The unique peptides for *B. thetaiotaomicron* are much higher in the Montana region with an average of 14.5. Examination of Figure 6.1, the national average for bacteria, illustrates that the Montana region has the highest average for all bacteria.

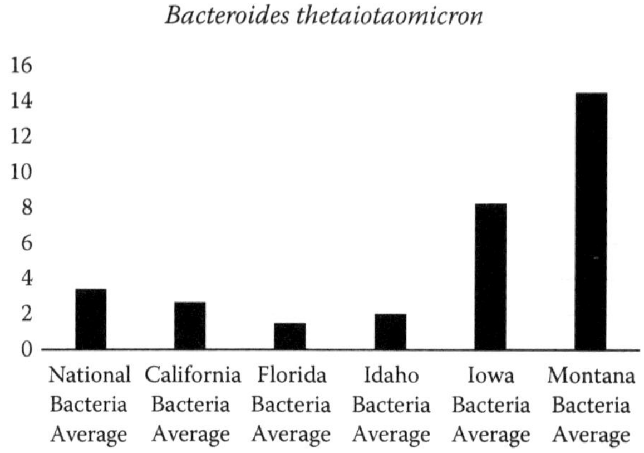

FIGURE 6.13 Detection and identification of *Bacteroides thetaiotaomicron* using the MSP/ABOid method.

6.2.13 *Burkholderia* sp.

Burkholderia are aerobic, gram-negative, nitrogen-fixing bacteria of the phylum Proteobacteria.

Burkholderia xenovorans is an important symbiont for plants, mostly grasses, as a nitrogen-fixing bacterium, particularly in nitrogen-poor soil. It is also important as a polychlorinated biphenyl (PCB) degrader.

The national average for *Burkholderia sp.* varies between species, the range is 3.0–4.6 unique peptides. The California average range is 2.4–4.3 and the Florida average is 0.9–1.8. The Idaho average is 1.8–4.3 unique peptides and Iowa has 5.6–9.0 unique peptides (Figure 6.14). The unique peptides for *Burkholderia sp.* are much higher in the Montana region with an average of 9.0–18. Examination of Figure 6.1, the national average for bacteria, illustrates that the Montana region has the highest average for all bacteria.

6.2.14 *Dehalococcoides ethenogenes*

Dehalococcoides ethenogenes is a gram-positive, motile, coccoid-shaped, anaerobic bacterium of the phylum Chloroflexi.

D. ethenogenes is the only known bacteria that can fully degrade tetrachloroethylene, a colorless chlorocarbon liquid widely used for the dry cleaning of fabrics, to ethene and as such has been used to treat groundwater to remove this compound. *Dehalococcoides* can also transform many highly toxic and/or persistent compounds including trichloroethene (TCE), vinyl chloride (VC), benzenes, PCBs, phenols and other similar compounds, making this microbe a useful means to remediate these compounds from the environment (Maphosa et al., 2012). *D. ethenogenes* can be expected to found where this activity has occurred.

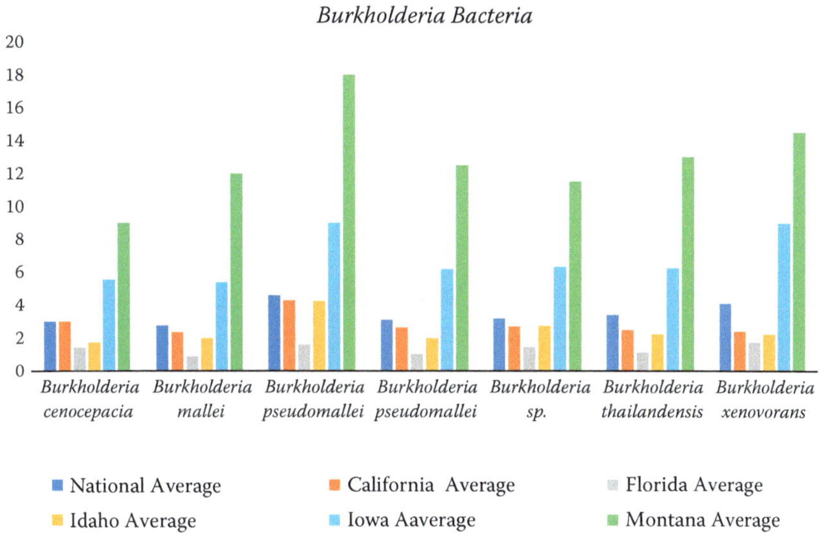

FIGURE 6.14 Detection and identification of six species of *Burkholderia* using the MSP/ABOid method.

The national average for *D. ethenogenes* is 3.2 unique peptides. The California average is 1.7 and the Florida average is 1.1. The Idaho average is 3.3 unique peptides and Iowa has 6.1 unique peptides (Figure 6.15). The unique peptides for *D. ethenogenes* are much higher in the Montana region with an average of 12.0. Examination of Figure 6.1, the national average for bacteria, illustrates that the Montana region has the highest average for all bacteria.

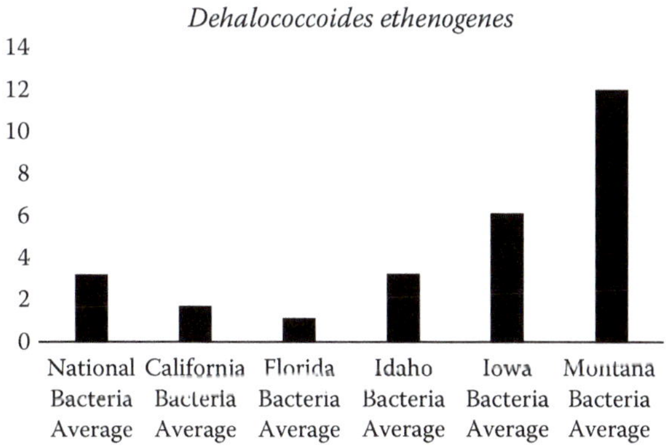

FIGURE 6.15 Detection and identification of *Dehalococcoides ethenogenes* using the MSP/ABOid method.

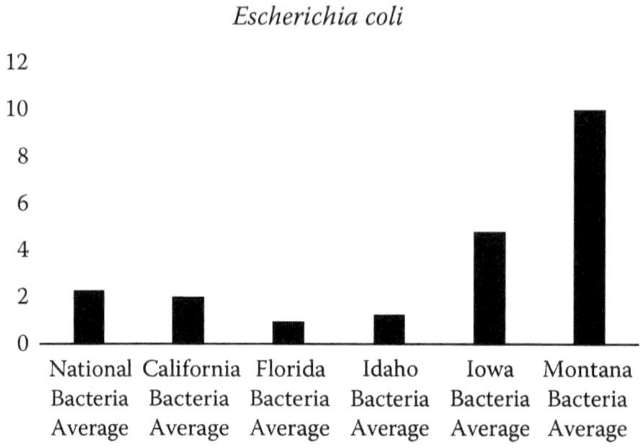

FIGURE 6.16 Detection and identification of *Escherichia coli* using the MSP/ABOid method.

6.2.15 *Escherichia coli*

Escherichia coli is a facultative anaerobic, gram-negative, non-sporulating, rod-shaped, coliform bacterium of the phylum Proteobacteria.

E. coli has been further classified into hundreds of strains on the basis of different serotypes, and is one of the most widely known bacterium and is commonly found in the lower intestines of mammals. As a result, a common use for *E. coli* is as a measure of water purification. In this use, the *E. coli* "index" indicates how much human feces is in the water or on commercial crops as they are being processed on food processing equipment and all similar uses. It is widely found in the environment and has even been isolated on the edge of hot springs.

The national average for *E. coli* is 2.3 unique peptides. The California average is 2.0 and the Florida average is 1.0. The Idaho average is 1.3 unique peptides and Iowa has 4.8 unique peptides (Figure 6.16). The unique peptides for *E.* are much higher in the Montana region with an average of 10.0. Examination of Figure 6.1, the national average for bacteria, illustrates that the Montana region has the highest average for all bacteria.

E. coli is so well known and identified with testing water and food that it is important to remember that there are many hundred strains of *E. coli* and what has been collected by the honeybee is only a small reference to this diversity. Other strains of *E. coli* are discussed in the chapter on water (Chapter 10). What is important to remember is that the honeybees collect a diverse population of microbes, and these are just a few that can be identified and reported.

6.2.16 *Frankia alni*

Frankia alni is a gram-positive, filamentous, non-motile bacterium of the phylum Actinobacteria. Some *Actinobacteria* can form rod- or coccoid- shaped forms, while

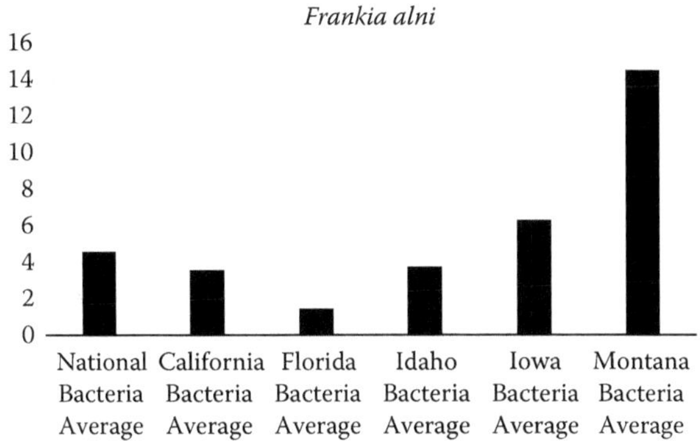

FIGURE 6.17 Detection and identification of *Frankia alni* using the MSP/ABOid method.

others can form spores on aerial hyphae that look like a mycelium and a branching growth pattern. *Actinobacteria* are diverse and contain a variety of subdivisions.

F. alni lives in symbiosis with actinorhizal plants in the genus *Alnus*. It is a nitrogen-fixing bacterium and forms nodules on the roots of alder trees and is widely distributed in the temperate regions of the northern hemisphere. Other species are found in Africa, and yet another is distributed in Central and South America.

The national average for *F. alni* is 4.6 unique peptides. The California average is 3.6 and the Florida average is 1.4. The Idaho average is 3.8 unique peptides and Iowa has 6.3 unique peptides (Figure 6.17). The unique peptides for *F. alni* are much higher in the Montana region with an average of 14.5. Examination of Figure 6.1, the national average for bacteria, illustrates that the Montana region has the highest average for all bacteria.

6.2.17 *Frankia sp.*

Frankia sp. are gram-positive, filamentous, non-motile, nitrogen-fixing bacteria of the phylum Actinobacteria. Some *Actinobacteria* can form rod- or coccoid-shaped forms, while others can form spores on aerial hyphae that look like a mycelium and a branching growth pattern. Actinobacteria are diverse and contain a variety of subdivisions. There are many species of *Frankia* and other than *F. alni* appear in enough numbers to be among the numerous bacteria picked up by the honeybees.

Of interest to the honeybee observations is that most *Frankia* strains are specific to a particular plant species. *Frankia* is also sensitive to its environment. This may be useful in identifying the area where honeybees are working.

The national average for *Frankia sp.* is 2.4 unique peptides. The California average is 2.2 and the Florida average is 1.0. The Idaho average is 2.0 unique peptides and Iowa has 3.8 unique peptides (Figure 6.18). The unique peptides for *Frankia sp.* are much higher in the Montana region with an average of 9.0.

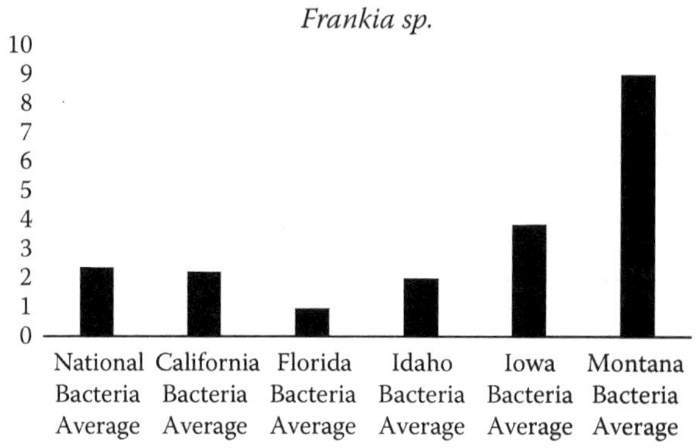

FIGURE 6.18 Detection and identification of *Frankia sp.* using the MSP/ABOid method.

Examination of Figure 6.1, the national average for bacteria, illustrates that the Montana region has the highest average for all bacteria.

6.2.18 *Geobacter sulfurreducens*

Geobacter sulfurreducens is a gram-negative, rod-shaped, obligately anaerobic, motile and metal- and sulphur-reducing bacterium of the phylum Proteobacterium.

G. sulfurreducens has the ability to produce electricity, which makes it an interesting candidate for useful fuel cells and electricity production in the future (Poddar & Khurana, 2011).

G. sulfurreducens has been isolated from water and soil contaminated by hydrocarbons.

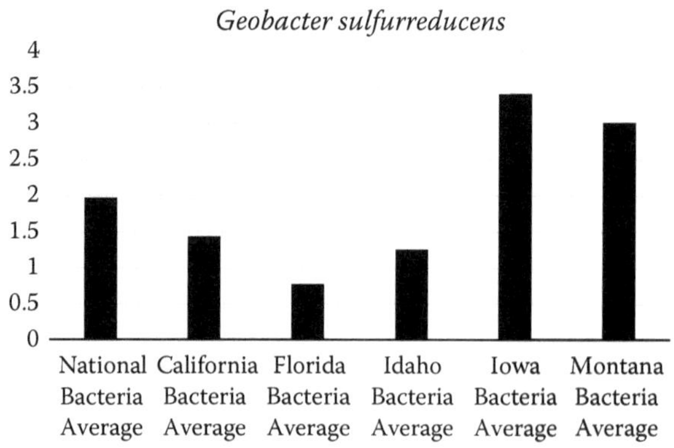

FIGURE 6.19 Detection and identification of *Geobacter sulfurreducens* using the MSP/ABOid method.

The national average for *G. sulfurreducens* is 2.0 unique peptides. The California average is 1.4 and the Florida average is 0.8. The Idaho average is 1.3 unique peptides and Iowa has 3.4 unique peptides, which is higher than the Montana region with an average of 3.0 (Figure 6.19). Examination of Figure 6.1, the national average for bacteria, illustrates the average for all bacteria.

6.2.19 *Hahella chejuensis*

Hahella chejuensis is a gram-negative, aerobic, rod-shaped and motile bacterium of the phylum Protobacteria. It was isolated from the the coastal marine sediment of the island of Marado, Korea (Lee et al., 2001).

H. chejuensis is a heterotrophic organism that produces a red pigment known to have lytic activity against *Cochlodinium polykrikoides*, a species of red tide-producing marine dinoflagellates.

The national average for *Hahella chejuensis* is 3.0 unique peptides. The California average is 2.4 and the Florida average is 1.1. The Idaho average is 2.0 unique peptides and Iowa has 6.0 unique peptides (Figure 6.20). The unique peptides for *Hahella chejuensis* are much higher in the Montana region with an average of 10.0. Examination of Figure 6.1, the national average for bacteria, illustrates that the Montana region has the highest average for all bacteria.

6.2.20 *Hyphomonas neptunium*

Hyphomonas neptunium is a heterotrophic, oligotroph, marine prostecate bacterium of the phylum Proteobacteria.

H. neptunium are able to grow in low concentrations of nutrients or even in areas that appear to have no nutrients. These prostecate and budding bacteria are found virtually everywhere, including some of the most extreme places like sea ice, hydrothermal vents and even in soils.

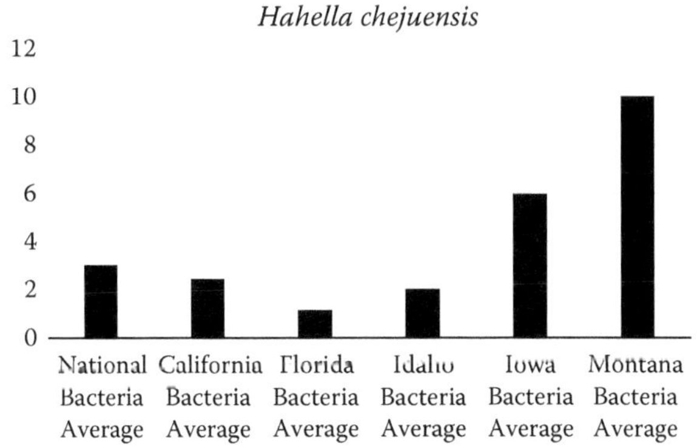

FIGURE 6.20 Detection and identification of *Hahella chejuensis* using the MSP/ABOid method.

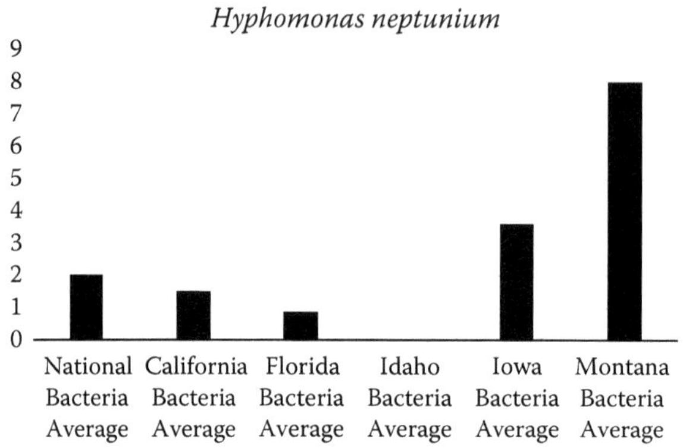

FIGURE 6.21 Detection and identification of *Hyphomonas neptunium* using the MSP/ABOid method.

The national average for *H. neptunium* is 2.0 unique peptides. The California average is 1.5 and the Florida average is 0.9. *H. neptunium* peptides are not found in the Idaho region. The Iowa region has 3.6 unique peptides (Figure 6.21). The unique peptides for *H. neptunium* are much higher in the Montana region with an average of 8.0. Examination of Figure 6.1, the national average for bacteria, illustrates that the Montana region has the highest average for all bacteria.

6.2.21 *Leptospira interrogans serovar*

Leptospira interrogans is a gram-negative, obligate aerobic, motile, and spiral- or helical-shaped bacterium of the phylum Spirochaetes.

L. interrogans causes leptospirosis, which occurs mostly in wild animals. Rodents are one of the main hosts of this bacterium, but it can infect domestic animals and even humans through contact with infected animal urine, either directly or in water. There are over 200 serotypes which are widely distributed. Because of the reservoir for this microbe in wild animals, likely sources of *L. interrogans* in the environment include sewers, stagnant water, farming and recreational water areas.

The national average for *L. interrogans* is 2.4 unique peptides. The California average is 1.7 and the Florida average is 0.8. *L. interrogans* peptides in the Idaho region was are 0.8. The Iowa region has 4.0 unique peptides (Figure 6.22). The unique peptides for *L. interrogans* are slightly higher in the Montana region with an average of 4.5. Examination of Figure 6.1, the national average for bacteria, illustrates that the Montana region has the highest average for all bacteria.

6.2.22 *Mesorhizobium loti*

Mesorhizobium loti, formerly known as *Rhizobium loti*, is a gram-negative, nitrogen-fixing, soil bacterium of the phylum Proteobacteria.

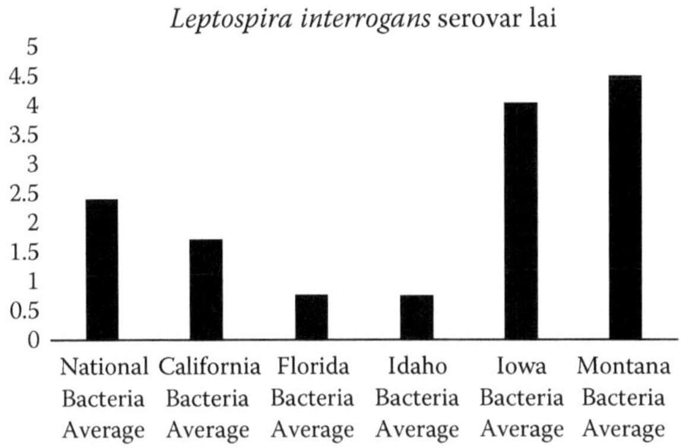

FIGURE 6.22 Detection and identification of *Leptospira interrogans* using the MSP/ABOid method.

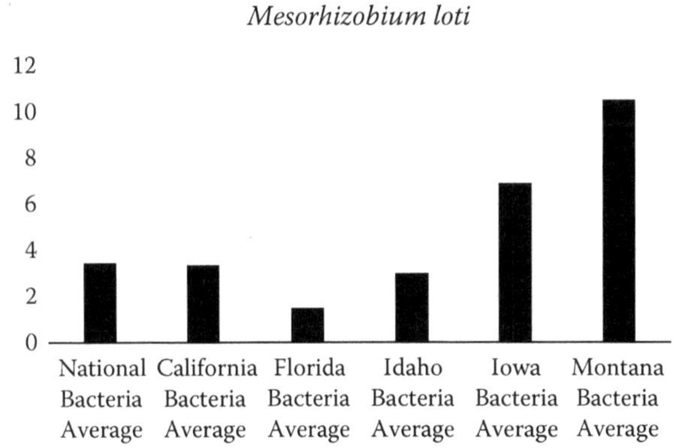

FIGURE 6.23 Detection and identification of *Mesorhizobium loti* using the MSP/ABOid method.

M. loti is found in the root nodules of many plant species.

The national average for *M. loti* is 3.4 unique peptides. The California average is 3.4 and the Florida average is 1.5. *M. loti* peptides in the Idaho region are 3.0. The Iowa region has 6.9 unique peptides (Figure 6.23). The unique peptides for *M. loti* are much higher in the Montana region with an average of 10.5. Examination of Figure 6.1, the national average for bacteria, illustrates that the Montana region has the highest average for all bacteria.

6.2.23 *Myxococcus xanthus*

Myxococcus xanthus is a gram-negative, rod-shaped species of myxobacteria with gliding motility of the phylum Proteobacteria.

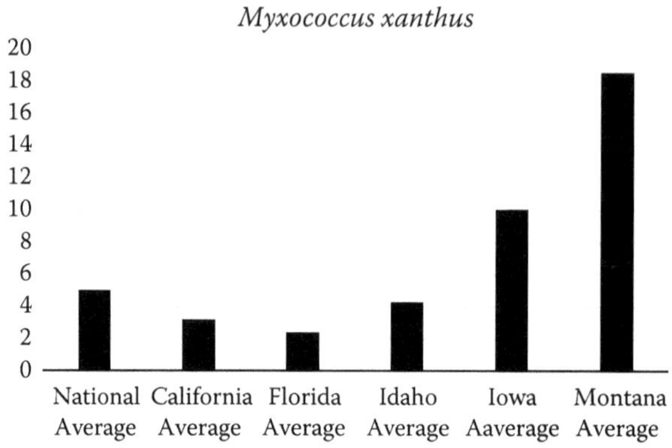

FIGURE 6.24 Detection and identification of *Myxococcus xanthus* using the MSP/ABOid method.

M. xanthus is a self-organized, saprotrophic and predatory soil bacterium that feeds on other bacteria.

The national average for *M. xanthus* is 5.0 unique peptides. The California average is 3.1 and the Florida average is 2.3. *M. xanthus* peptides in the Idaho region are 4.3. The Iowa region has 10.0 unique peptides (Figure 6.24). The unique peptides for *M. xanthus* are much higher in the Montana region with an average of 18.5. Examination of Figure 6.1, the national average for bacteria, illustrates that the Montana region has the highest average for all bacteria.

6.2.24 *Nitrobacter hamburgensis*

Nitrobacter hamburgensis is a gram-negative, rod- to pear-shaped, motile, obligate aerobic, nitrite-oxidizing bacteria of the phylum Proteobacteria (Bock, Sundermeyer-Klinger, & Stackebrandt, 1983).

N. hamburgensis has been useful in removing high levels of nitrogen from municipal effluents in wastewater. It is a soil bacterium that is also found in fresh water.

The national average for *N. hamburgensis* is 2.6 unique peptides. The California average is 2.4 and the Florida average is 1.1. The Idaho average is 1.8 unique peptides and Iowa has 5.2 unique peptides (Figure 6.25). The unique peptides for *N. hamburgensis* are about the same in the Montana region with an average of 5.0. Examination of Figure 6.1, the national average for bacteria, illustrates the average for all bacteria.

6.2.25 *Nocardia farcinica*

Nocardia farcinica is a gram-positive, rod-shaped bacterium of the phylum Actinobacteria.

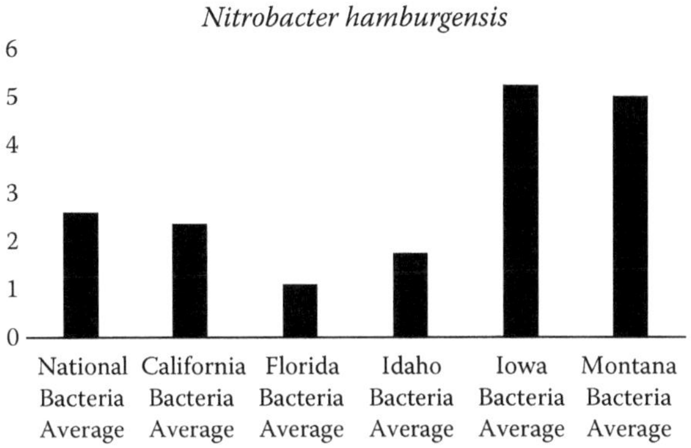

FIGURE 6.25 Detection and identification of *Nitrobacter hamburgensis* using the MSP/ABOid method.

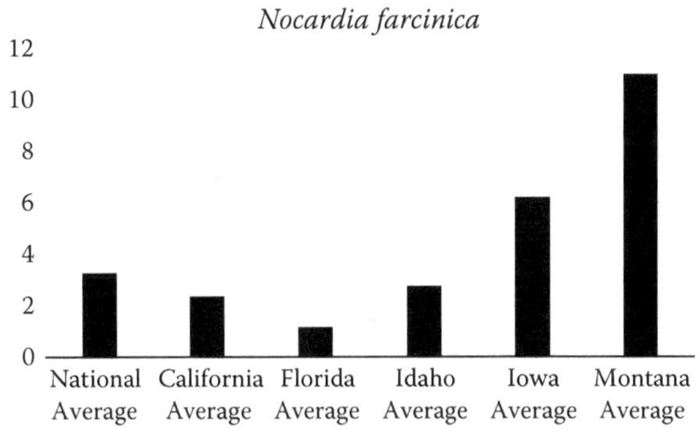

FIGURE 6.26 Detection and identification of *Nocardia farcinica* using the MSP/ABOid method.

N. farcinica is saprophytic and frequents soils, plants, animal tissue and other sources such as dead or decaying organic materials.

The national average for *N. farcinica* is 3.3 unique peptides. The California average is 2.4 and the Florida average is 1.1. The Idaho average is 2.8 unique peptides and Iowa has 6.2 unique peptides (Figure 6.26). The unique peptides for *N. farcinica* are greater in the Montana region with an average of 11.0. Examination of Figure 6.1, the national average for bacteria, illustrates the average for all bacteria.

6.2.26 *Nostoc* sp.

Nostoc sp. are photosynthesizers that use cytoplasmic photosynthetic pigments rather than chloroplasts for metabolism, motile via a swaying motion (lacking flagella), nitrogen-fixing bacteria of the phylum Cyanobacteria.

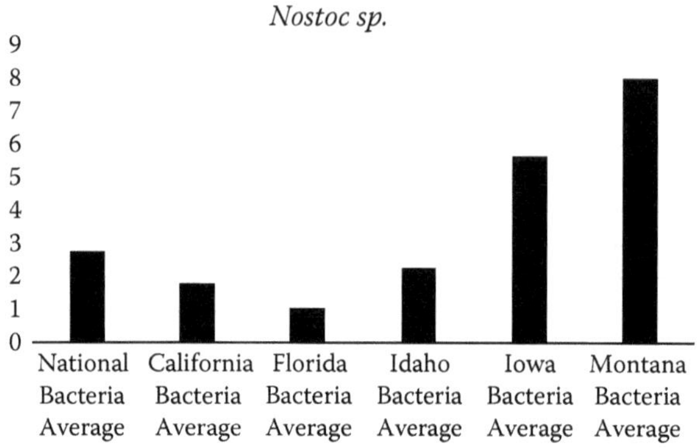

FIGURE 6.27 Detection and identification of *Nostoc* sp. using the MSP/ABOid method.

Nostoc sp. is a diverse genus known for its ability to withstand freezing and thawing cycles and to lie dormant for years and able to recover when rehydrated with water. They are distributed worldwide and one of the most abundant phototrophic bacteria capable of surviving in extreme environments. Some species can survive even in areas of extreme UV radiation.

Nostoc sp. have been found in fresh water, soils, on moist rocks and in extremely cold and extremely arid habitats. Their ability to fix nitrogen has allowed a niche in symbiotic interactions with organisms including fungi, lichens, mosses and ferns (Dodds, 1995).

The national average for *Nostoc* sp. is 2.8 unique peptides. The California average is 1.8 and the Florida average is 1.0. The Idaho average is 2.3 unique peptides and Iowa has 5.6 unique peptides. The unique peptides for *Nostoc* sp. are about the same in the Montana region with an average of 8.0 (Figure 6.27). Examination of Figure 6.1, the national average for bacteria, illustrates the average for all bacteria.

6.2.27 *Polaromonas* sp.

Polaromonas is a genus of gram-negative bacteria from the phylum Proteobacteria. *Polaromonas* species are psychrophiles. These are extremophilic organisms capable of growth and reproduction in low temperatures, ranging from −20°C to +10°C.

The national average for *Polaromonas* sp. is 2.5 unique peptides. The California average is 1.8 and the Florida average is 1.1. The Idaho average is 2.8 unique peptides and Iowa has 4.5 unique peptides. The unique peptides for *Polaromonas* sp. are much higher in the Montana region with an average of 12.5 (Figure 6.28). Examination of Figure 6.1, the national average for bacteria, illustrates the average for all bacteria.

6.2.28 *Pseudomonas aeruginosa*

Pseudomonas aeruginosa is a common, encapsulated, gram-negative, rod-shaped, motile, facultative anaerobic bacterium of the phylum Proteobacteria.

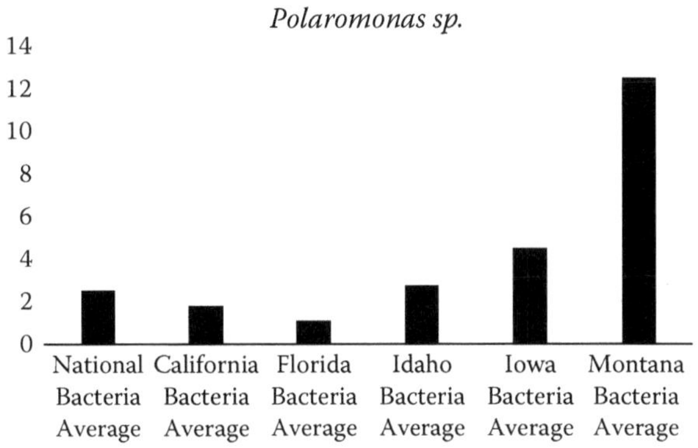

FIGURE 6.28 Detection and identification of *Polaromonas sp.* using the MSP/ABOid method.

It is found in soil, water, skin flora and most man-made moist environments such as hot tubs and swimming pools throughout the world. *P. aeruginosa* is an opportunistic pathogen and responsible for infections in plants, animals (including invertebrates) and insects such as the fruit fly and moths.

P. aeruginosa is a ubiquitous microorganism and may be the most abundant organism on earth as a result of its ability to grow in all aquatic ecosystems where it is the predominant inhabitant.

The national average for *P. aeruginosa* is 3.4 unique peptides. The California average is 2.4 and the Florida average is 1.1. The Idaho average is 2.8 unique peptides and Iowa has 5.0 unique peptides. The unique peptides for *P. aeruginosa* are much higher in the Montana region with an average of 8.5 (Figure 6.29). Examination of Figure 6.1, the national average for bacteria, illustrates the average for all bacteria.

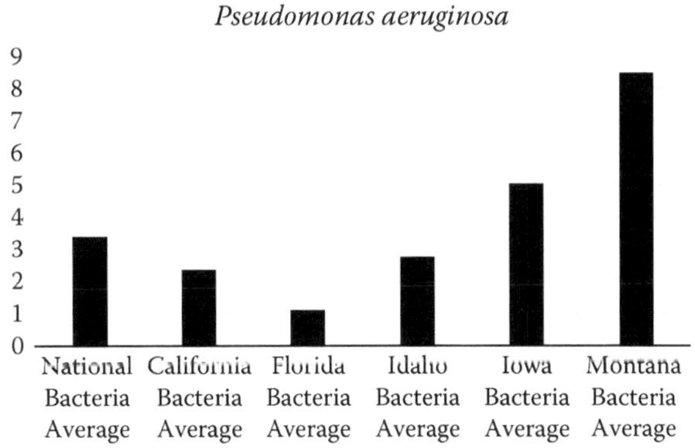

FIGURE 6.29 Detection and identification of *Pseudomonas aeruginosa* using the MSP/ABOid method.

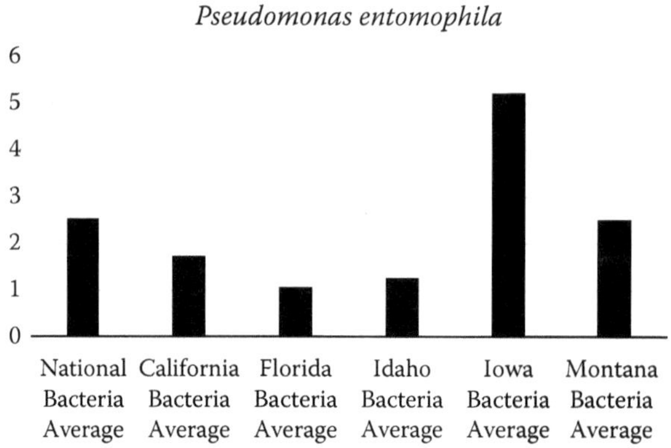

FIGURE 6.30 Detection and identification of *Pseudomonas entomophila* using the MSP/ABOid method.

6.2.29 *Pseudomonas entomophila*

Pseudomonas entomophila is a gram-negative, rod-shaped, motile, facultative anaerobic bacterium of the phylum Proteobacteria.

P. entomophila are diverse and found in soil, aquatic and rhizosphere environments. This species is highly toxic to insects, but not to plants, and is closely related to *Pseudomonas putida* (Vodovar et al., 2006).

The national average for *P. entomophila* is 2.5 unique peptides. The California average is 1.7 and the Florida average is 1.0. The Idaho average is 1.3 unique peptides and Iowa has 5.2 unique peptides. The unique peptides for *P. entomophila* are lower in the Montana region with an average of 2.5 (Figure 6.30). Examination of Figure 6.1, the national average for bacteria, illustrates the average for all bacteria.

6.2.30 *Pseudomonas fluorescens*

Pseudomonas fluorescens is a common, gram-negative, motile, facultative anaerobic bacterium of the phylum Proteobacteria.

The national average for *P. fluorescens* is 2.7 unique peptides. The California average is 2.4 and the Florida average is 1.2. The Idaho average is 2.8 unique peptides and Iowa has 5.2 unique peptides. The unique peptides for *P. fluorescens* are higher in the Montana region with an average of 10.5 (Figure 6.31). Examination of Figure 6.1, the national average for bacteria, illustrates the average for all bacteria.

6.2.31 *Pseudomonas putida*

Pseudomonas putida is a gram-negative, rod-shaped, motile, multiplasmid, hydrocarbon-degrading, aerobic, saprotrophic soil bacterium of the phylum Proteobacteria.

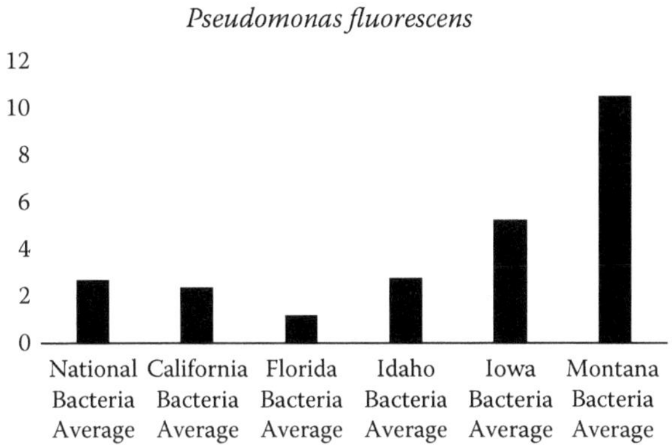

FIGURE 6.31 Detection and identification of *Pseudomonas fluorescens* using the MSP/ABOid method.

P. putida is commonly found in soil, plants and water, where it produces a variety of secondary metabolites including antibiotics against soil-borne plant pathogens.

P. putida induces plant growth and protects the plants from pathogens, and as a result of a diverse metabolism that can degrade organic solvents, such as toluene, through oxidative reactions, it has become a useful microbe in bioremediation.

The national average for *P. putida* is 2.9 unique peptides. The California average is 1.5 and the Florida average is 1.0. The Idaho average is 1.3 unique peptides and Iowa has 5.5 unique peptides. The unique peptides for *P. putida* are higher in the Montana region with an average of 7.5 (Figure 6.32). Examination of Figure 6.1, the national average for bacteria, illustrates the average for all bacteria.

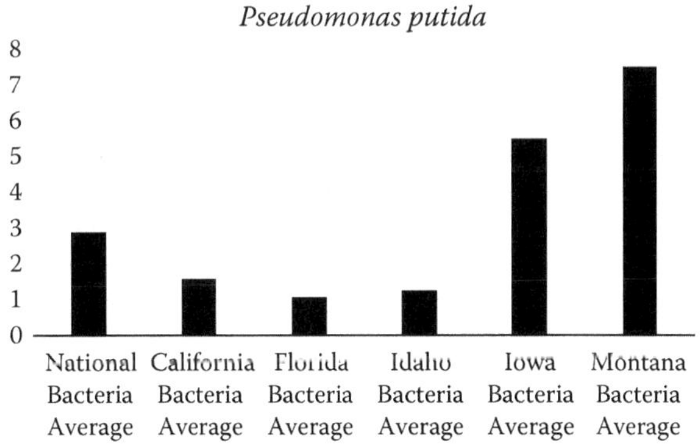

FIGURE 6.32 Detection and identification of *Pseudomonas putida* using the MSP/ABOid method.

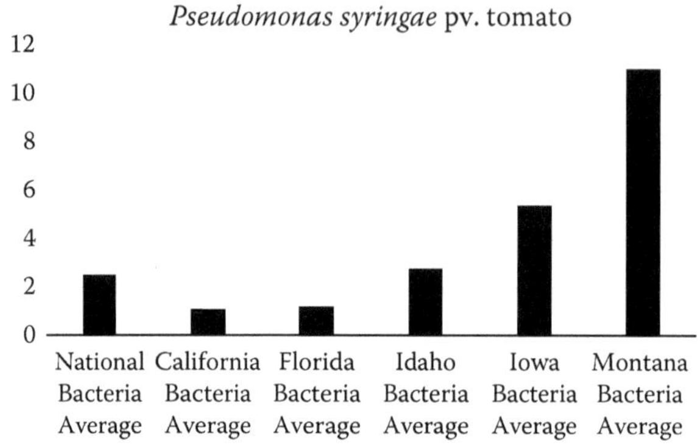

FIGURE 6.33 Detection and identification of *Pseudomonas syringae* using the MSP/ABOid method.

6.2.32 *Pseudomonas syringae pv. Tomato*

Pseudomonas syringae pv. tomato (*Pseudomonas tomato*) is a gram-negative, rod-shaped, motile bacterium of the phylum Proteobacteria.

Pseudomonas tomato is a pathogen that infects a variety of plants including the tomato.

The national average for *Pseudomonas syringae pv. tomato* is 2.5 unique peptides. The California average is 1.1 and the Florida average is 1.2. The Idaho average is 2.8 unique peptides and Iowa has 5.4 unique peptides. The unique peptides for *Pseudomonas syringae pv. tomato* are higher in the Montana region with an average of 11.0 (Figure 6.33). Examination of Figure 6.1, the national average for bacteria, illustrates the average for all bacteria.

6.2.33 *Ralstonia eutropha*

Ralstonia eutropha is a gram-negative, rod-shaped, non-spore-forming, motile, facultative aerobic bacterium of the phylum Proteobacteria.

R. eutropha is found in multiple inhabitants, but usually in both soil and water environments, where it thrives in the presence of millimolar concentrations of heavy metals, including zinc, cadmium, cobalt, lead, mercury, nickel and chromium.

R. eutropha is useful in bioremediation as it is able to degrade numerous chlorinated aromatic (chloroaromatic) compounds.

The national average for *R. eutropha* is 3.6 unique peptides. The California average is 2.4 and the Florida average is 1.2. The Idaho average is 1.8 unique peptides and Iowa has 6.9 unique peptides. The unique peptides for *R. eutropha* are higher in the Montana region with an average of 14.5 (Figure 6.34). Examination of Figure 6.1, the national average for bacteria, illustrates the average for all bacteria.

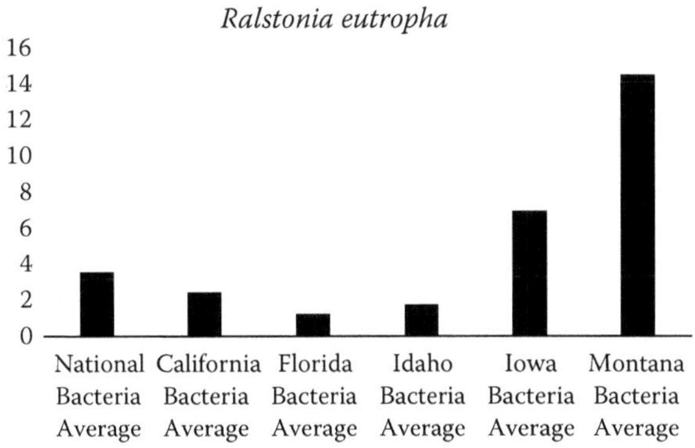

FIGURE 6.34 Detection and identification of *Ralstonia eutropha* using the MSP/ABOid method.

6.2.34 *Rhodococcus sp.*

Rhodococcus is a genus of gram-positive, non-motile, filamentous, rod-forming, non-sporulating, aerobic bacteria of the phylum Actinobacteria.

Rhodococcus bacteria reside in soil and water habitats, where they are considered one of "the most industrial important organisms" as a result of their ability in the biodesulfurization of fossil fuels, degradation of PCBs and their utilization of a wide variety of other organic compounds as energy sources (Lichtinger, Reiss, and Benz, 2000).

The national average for *Rhodococcus sp.* is 4.4 unique peptides. The California average is 3.1 and the Florida average is 1.9. The Idaho average is 3.5 unique peptides and Iowa has 8.6 unique peptides. The unique peptides for *Rhodococcus sp.* are higher in the Montana region with an average of 13.0 (Figure 6.35). Examination of Figure 6.1, the national average for bacteria, illustrates the average for all bacteria.

6.2.35 *Rhodoferax ferrireducens*

Rhodoferax ferrireducens is a psychrotolerant, facultative anaerobic, gram-negative, rod-shaped, motile, non-spore-forming bacterium of the phylum Proteobacteria.

R. ferrireducens was first isolated in the sediments in the waters around Oyster Bay, VA (United States) and are frequently found in stagnant aquatic systems exposed to light (Imhoff, 2006). Other environments include pond water, ditch water and sewage.

R. ferrireducens can produce electricity as it feeds on sugars (Graham, 2003), making it a candidate as a microbial fuel cell because this organism cannot only oxidize carbohydrates but also other metabolic intermediates such as acetate, lactate and pyruvate.

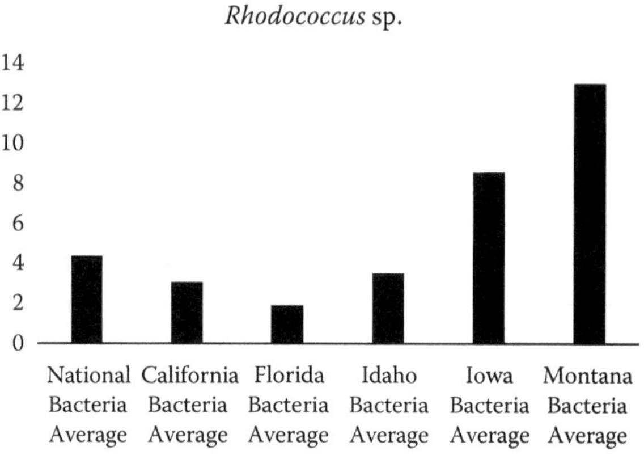

FIGURE 6.35 Detection and identification of *Rhodococcus* using the MSP/ABOid method.

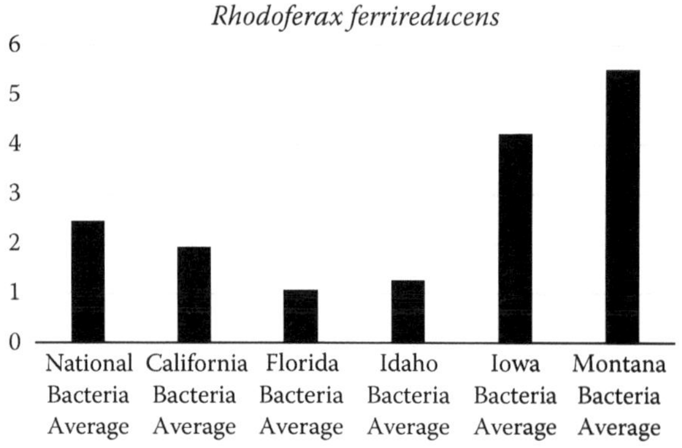

FIGURE 6.36 Detection and identification of *Rhodoferax ferrireducens* the MSP/ABOid method.

The national average for *R. ferrireducens* is 2.4 unique peptides. The California average is 1.9 and the Florida average is 1.0. The Idaho average is 1.3 unique peptides and Iowa has 4.2 unique peptides. The unique peptides for *R. ferrireducens* are higher in the Montana region with an average of 5.5 (Figure 6.36). Examination of Figure 6.1, the national average for bacteria, illustrates the average for all bacteria.

6.2.36 *Rhodopirellula baltica*

Rhodopirellula baltica is a marine bacterium that can be found all over the world in terrestrial and marine habitats alike.

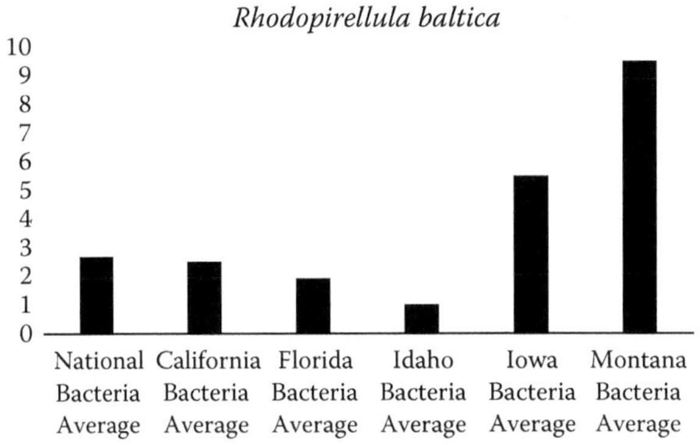

FIGURE 6.37 Detection and identification of *Rhodopirellula baltica* using the MSP/ABOid method.

The national average for *R. baltica* is 2.7 unique peptides. The California average is 2.5 and the Florida average is 1.9. The Idaho average is 1.0 unique peptide and Iowa has 5.5 unique peptides. The unique peptides for *R. baltica* are higher in the Montana region with an average of 9.5 (Figure 6.37). Examination of Figure 6.1, the national average for bacteria, illustrates the average for all bacteria.

6.2.37 *Rhodopseudomonas palustris*

Rhodopseudomonas palustris is a rod-shaped, gram-negative, motile, purple, non-sulfur bacterium, able to grow with or without oxygen, of the phylum Proteobacteria

R. palustris can flexibly switch among any of the four modes of metabolism that support life making it be found extensively in nature. It has, among other sources, been isolated from swine waste lagoons, earthworm droppings, marine coastal sediments and pond water.

R. palustris can metabolize lignin and acids found in degrading plant and animal waste by metabolizing carbon dioxide and aromatic compounds, making it an efficient biodegradation catalyst in both aerobic and anaerobic environments.

The national average for *R. palustris* is 3.2 unique peptides. The California average is 2.8 and the Florida average is 1.9. The Idaho average is 1.8 unique peptides and Iowa has 5.0 unique peptides. The unique peptides for *R. palustris* are higher in the Montana region with an average of 10.5 (Figure 6.38). Examination of Figure 6.1, the national average for bacteria, illustrates the average for all bacteria.

6.2.38 *Salmonella typhimurium*

Salmonella typhimurium is a pathogenic, gram-negative bacterium of the Phylum Proteobacteria.

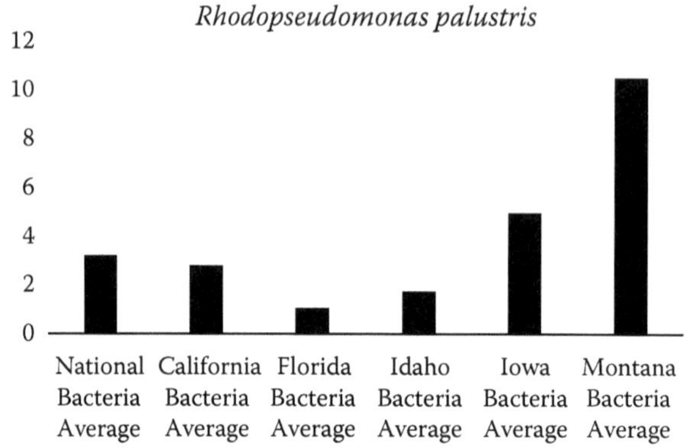

FIGURE 6.38 Detection and identification of *Rhodopseudomonas palustris* using the MSP/ABOid method.

Salmonella bacteria are common on raw egg shells, in poultry and red meat, and are part of natural bacterial flora of reptiles and amphibians. Contaminated water is one of the major sources of *Salmonella* infections worldwide.

The national average for *S. typhimurium* is 1.9 unique peptides. The California average is 2.1 and the Florida average is 1.0. The Idaho average is 2.3 unique peptides and Iowa has 5.0 unique peptides. The unique peptides for *S. typhimurium* are higher in the Montana region with an average of 6.0 (Figure 6.39). Examination of Figure 6.1, the national average for bacteria, illustrates the average for all bacteria.

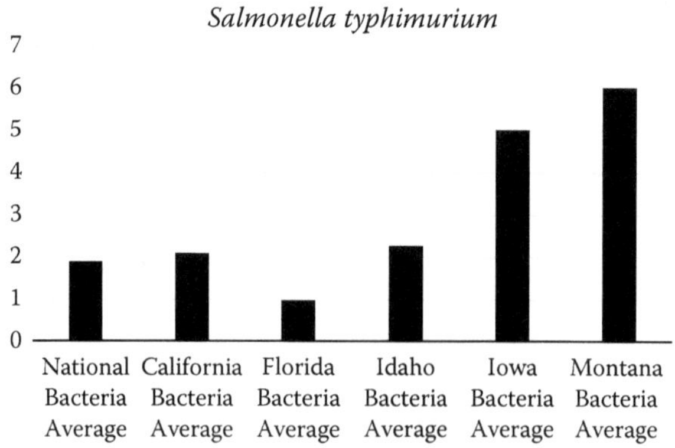

FIGURE 6.39 Detection and identification of *Salmonella typhimurium* using the MSP/ABOid method.

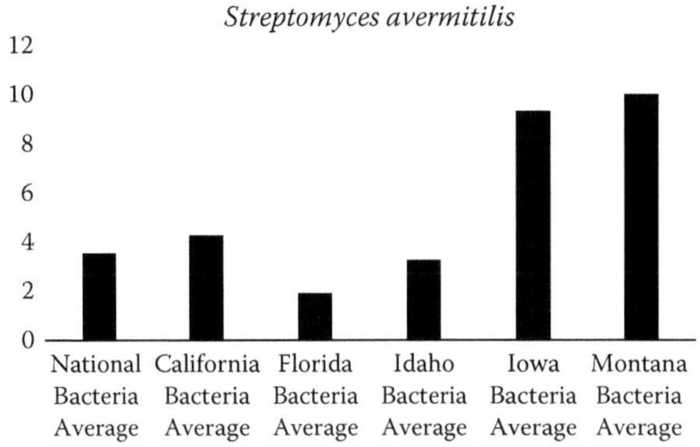

FIGURE 6.40 Detection and identification of *Streptomyces avermitilis* using the MSP/ABOid method.

6.2.39 *Streptomyces avermitilis*

Streptomyces avermitilis is a spore-forming, aerobic, hyphae-producing, gram-positive bacterium of the phylum Actinobacteria. There are over 500 species described.

S. avermitilis is found in soils where they are helpful to other soil microbes by killing nearby nematodes by the biosynthesis of anthelmintics. *S. avermitilis* also has insecticidal properties.

The national average for *S. avermitilis* is 3.6 unique peptides. The California average is 4.3 and the Florida average is 1.9. The Idaho average is 1.8 unique peptides and Iowa has 9.3 unique peptides. The unique peptides for *S. avermitilis* are higher in the Montana region with an average of 10.0 (Figure 6.40). Examination of Figure 6.1, the national average for bacteria, illustrates the average for all bacteria.

6.2.40 *Streptomyces coelicolor*

Streptomyces coelicolor is a spore-forming, aerobic, hyphae-producing, gram-positive bacterium of the phylum Actinobacteria. There are over 500 species described.

S. coelicolor, like the *Streptomyces* genus in general, live in the soil where they are capable of living on many different carbon sources. *Streptomyces* are responsible for much of the breakdown of organic material and are known for the "earthy" smell of soil.

The national average for *S. coelicolor* is 4.6 unique peptides. The California average is 3.4 and the Florida average is 2.4. The Idaho average is 3.8 unique peptides and Iowa has 9.0 unique peptides. The unique peptides for *S. coelicolor* are higher in the Montana region with an average of 16.0 (Figure 6.41). Examination of Figure 6.1, the national average for bacteria, illustrates the average for all bacteria.

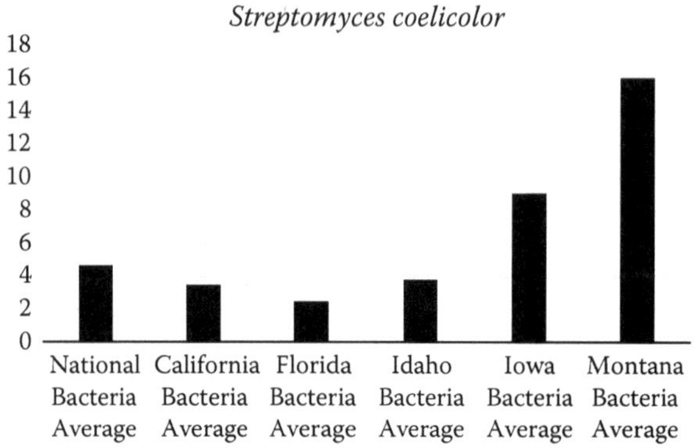

FIGURE 6.41 Detection and identification of *Streptomyces coelicolor* using the MSP/ABOid method.

6.2.41 *Trichodesmium erythraeum*

Trichodesmium erythraeum is a gram-negative, motile, nitrogen and carbon-fixing, photosynthesis-capable, filamentatious bacterium of the phylum Cyanobacteria.

Trichodesmium is found in tropical and subtropical ocean waters with low nutrient levels.

T. erythraeum is a carbon and nitrogen fixing bacterium, and like other Cyanobacteria, *T. erythraeum* is able to derive energy through the process of photosynthesis.

Trichodesmium species is important to the global ecosystem because it contributes upwards of 40% of all nitrogen fixing occurring in the ocean (Karl et al., 2002).

The national average for *T. erythraeum* is 2.2 unique peptides. The California average is 2.1 and the Florida average is 0.7. The Idaho average is 1.8 unique

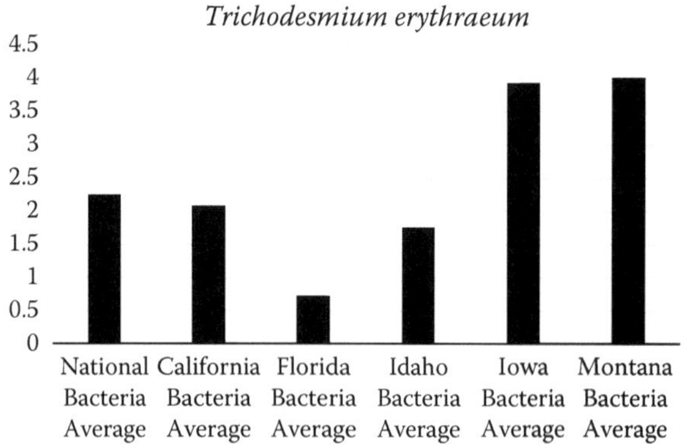

FIGURE 6.42 Detection and identification of *Trichodesmium erythraeum* using the MSP/ABOid method.

peptides and Iowa has 3.9 unique peptides. The unique peptides for *T. erythraeum* are similar in the Montana region with an average of 4.0 (Figure 6.42). Examination of Figure 6.1, the national average for bacteria, illustrates the average for all bacteria.

6.2.42 *Vibrio vulnificus*

Vibrio vulnificus is a gram-negative, curved, rod-shaped, motile, halophilic bacterium of the phylum Proteobacteria.

V. vulnificus is a virulent bacterium typically found in salty coastal waters thriving especially in molluscan shellfish including oysters and clams.

The national average for *V. vulnificus* is 2.0 unique peptides. The California average is 1.9 and the Florida average is 0.7. The Idaho average is 1.0 unique peptides and Iowa has 3.3 unique peptides. The unique peptides for *V. vulnificus* are higher in the Montana region with an average of 6.0 (Figure 6.43). Examination of Figure 6.1, the national average for bacteria, illustrates the average for all bacteria.

6.2.43 *Xanthomonas axonopodis* pv. *Citri*

Xanthomonas axonopodis is a gram-negative, non-spore-forming, motile, aerobic, rod-shaped bacterium of the phylum Proteobacteria.

Xanthomonas citri is part of the *Xanthomonas* genus of phytopathogenic bacteria that inhabit the phyllosphere of citrus plants around the world and is most known to cause the citrus canker disease.

The national average for *X. axonopodis* is 3.2 unique peptides. The California average is 2.8 and the Florida average is 1.9. The Idaho average is 1.8 unique peptides and Iowa has 5.0 unique peptides. The unique peptides for *X. axonopodis* are higher in the Montana region with an average of 10.5 (Figure 6.44). Examination of Figure 6.1, the national average for bacteria, illustrates the average for all bacteria.

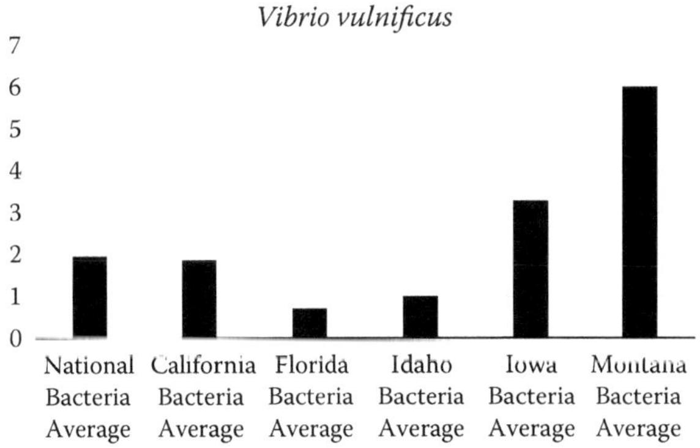

FIGURE 6.43 Detection and identification of *Vibrio vulnificus* using the MSP/ABOid method.

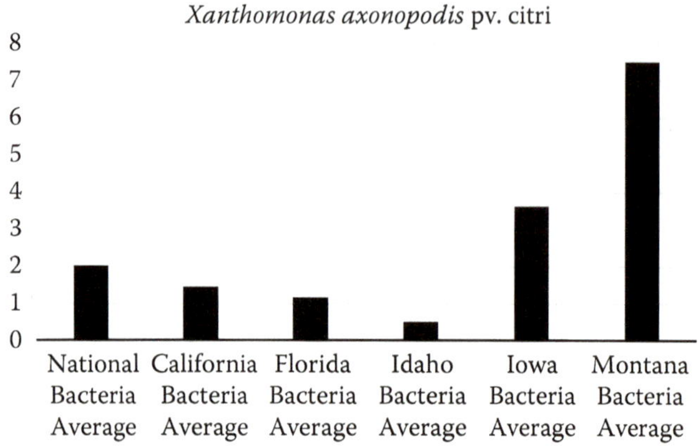

FIGURE 6.44 Detection and identification of *Xanthomonas axonopodis* using the MSP/ABOid method.

6.3 CALIFORNIA REGIONAL AVERAGE

Figure 6.45 illustrates the distribution of bacteria in the California region. The averages for more than 200 bacteria range from less than 0.5 to 5.0 unique

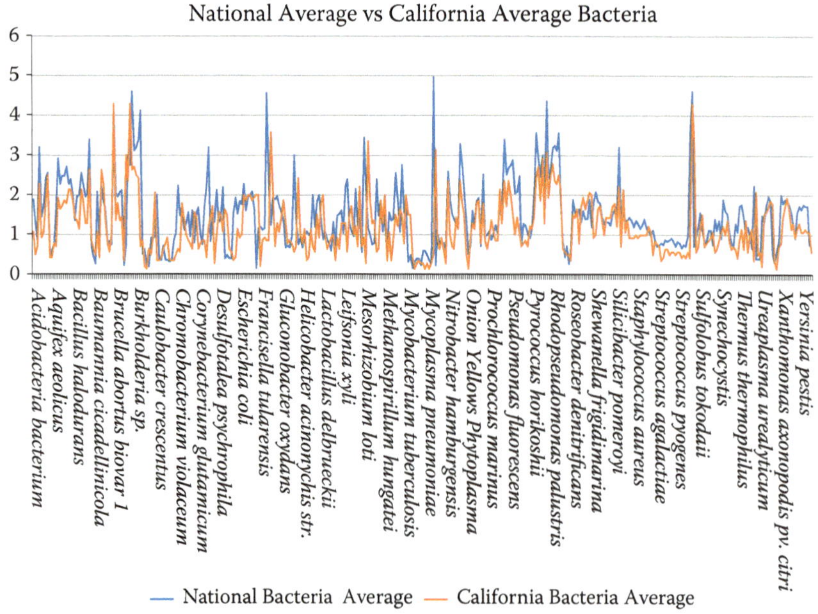

FIGURE 6.45 National and unique bacteria peptide averages for California regional average using the MSP/ABOid method. Listing of all California bacteria and their average of unique peptides is given in Appendix B.

peptides. Details are provided in Appendix B, which has all the bacteria and their respective average.

6.4 FLORIDA REGIONAL AVERAGE

Figure 6.46 illustrates the distribution of bacteria in the Florida region. The averages for more than 200 bacteria range from less than 0.5 to 3.5 unique peptides. Details are provided in Appendix C, which has all the bacteria and their respective average.

6.5 IDAHO REGIONAL AVERAGE

Figure 6.47 illustrates the distribution of bacteria in the Idaho region. The averages for more than 200 bacteria range from less than 0.5 to 4.1 unique peptides. Details are provided in Appendix D, which has all the bacteria and their respective averages.

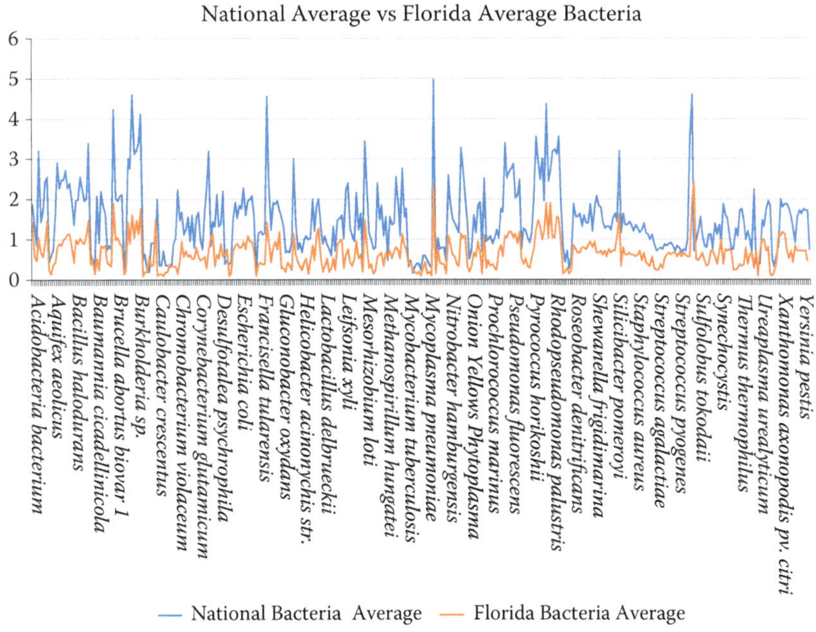

FIGURE 6.46 National and unique bacteria peptide averages for Florida regional average using the MSP/ABOid method. Listing of all Florida bacteria and their average of unique peptides is given in Appendix C.

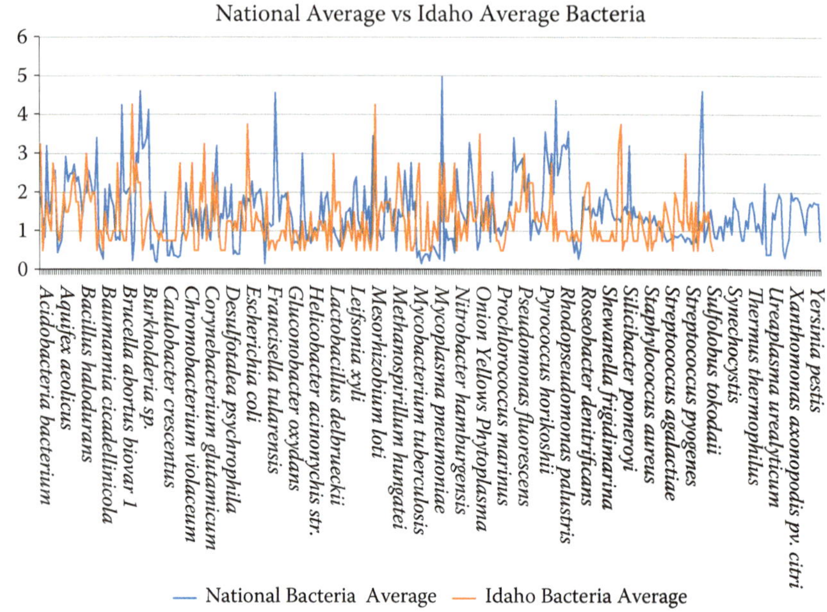

FIGURE 6.47 National and unique bacteria peptide averages for Idaho regional average using the MSP/ABOid method. Listing of all Idaho bacteria and their average of unique peptides is given in Appendix D.

6.6 IOWA REGIONAL AVERAGE

Figure 6.48 illustrates the distribution of bacteria in the Iowa region. The averages for more than 200 bacteria range from less than 0.5 to 10.0 unique peptides. Details are provided in Appendix E, which has all the bacteria and their respective average.

6.7 MONTANA REGIONAL AVERAGE

Figure 6.49 illustrates the distribution of bacteria in the Montana region. The averages for more than 200 bacteria range from less than 0.5 to 3.5 unique peptides. Details are provided in Appendix F, which has all the bacteria and their respective average.

6.8 NATIONAL AND REGIONAL AVERAGES

Figure 6.50 illustrates the distribution of bacteria for all the regions and the national average. These averages are for more than 200 bacteria and range from

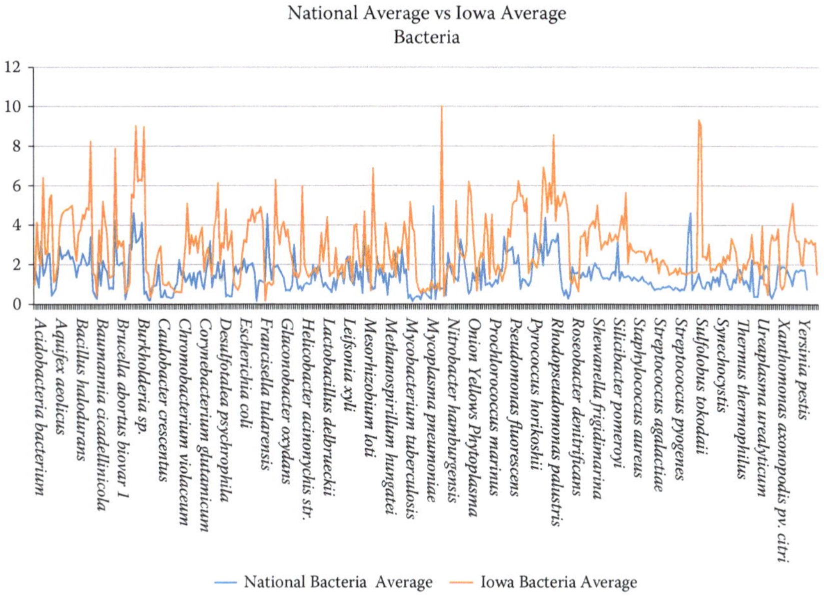

FIGURE 6.48 National and unique bacteria peptide averages for Iowa regional average using the MSP/ABOid method. Listing of all Iowa bacteria and their average of unique peptides is given in Appendix E.

less than 0.5 to 18.0 unique peptides. Details of the bacteria species and their averages are provided in Appendices A–F, which have all the bacteria and their respective average.

6.9 DISCUSSION

Several hundred bacteria have been identified. Many are bacteria with symbiotic relationships with plants, and others are soil inhabitants. The national average for these bacteria ranges from less than 1 to 5 unique peptides. Around 40 are discussed in detail, and of these 8 are the highest averaging from 3 to 5 unique peptides. These are further classified as three bacteria having a relationship with symbiotic mycorrhizal fungi, two being bioremediation bacteria and the others being soil inhabitants including *Myxococcus*, a gliding bacterium that feeds on microorganisms including bacteria and fungi.

The classification of bacteria is ongoing and some of them may be underspeciated at the moment because there are many unsequenced and un-named bacteria. The rapid increase in sequencing bacteria will contribute to identification and cement phylogenic relationships. Nevertheless, in this chapter hundreds of bacteria have been detected and discussed.

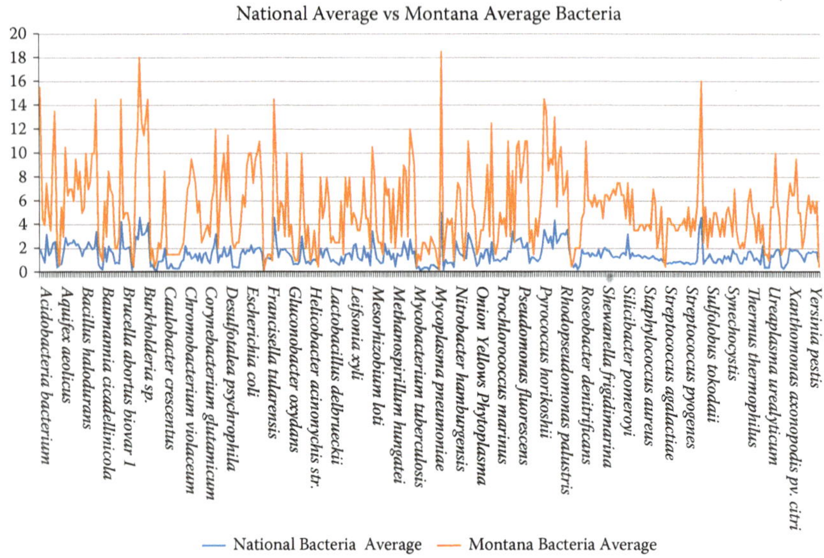

FIGURE 6.49 National and unique bacteria peptide averages for Montana regional average using the MSP/ABOid method. Listing of all Montana bacteria and their average of unique peptides is given in Appendix F.

Environmentalists are encouraged to study these bacteria and compare them with the fungi and viruses to determine mutual relationships and assess these relationships within a region, area or a locality .

The number of bacteria species that can be identified is continuing to increase due to the increasing number of bacterial sequences added to the NCBI. More than 250,000 prokaryotic microbes have been sequenced but only 19,379 bacteria have complete sequences. This number increases daily.

The ability to add new microbes has been discussed in Chapter 2, but it remains important to remember that as new bacteria are sequenced and added to the Bacteria Data Group the samples that have been analyzed can be re-analyzed using the old computer files. In this manner there is no need to collect a new sample and large archived sample sets can be re-examined and a search made for the new bacteria.

Several bacteria stand out from the others, such as *Burkholderia*, *Frankia*, *Streptomyces* and *Pseudomonas*. There are others that show increases or decreases among the regions, but these are much higher than the national averages.

Finally, considering all the bacteria species that exist and are being discovered, it is often possible to consider that we live in a very rich environment of microbes of which bacteria are a large part. There is so much potential for this

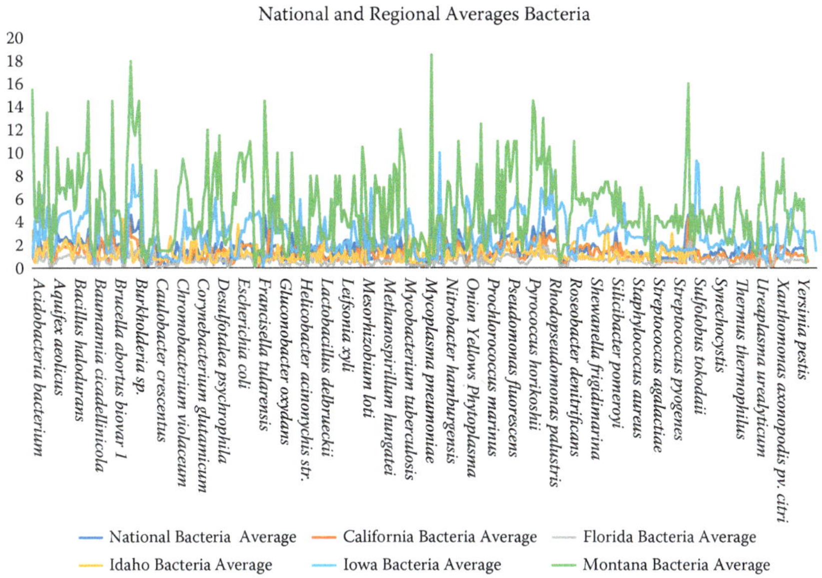

FIGURE 6.50 National and regional averages for unique bacteria peptides.

type of information collected by the honeybees. In their regular work they have the information about their working environment with them, all we have to do is sample, detect and identify the microbes. Microbes of all sorts are found are with the honeybees and the diversity of these microbes may be useful in measuring or monitoring changes to the environment.

7 Fungi

Fungi along with bacteria and viruses are among the most numerous of microbes. It can be expected that the honeybee would collect fungi during foraging.

The national average for fungi is presented first for all 200 fungi detected. From this list 25 of the 40 most frequent fungi determined as having an average of over 13 unique peptides are discussed in greater detail and compared with the averages for five regions: California, Florida, Idaho, Iowa and Montana. To simplify these frequent fungi, description are provided that include taxonomic information to phylum, a description of where they are likely to be found in nature, a chart showing the relationship with the national average and the average of the five regions and a summary of that relationship.

It is helpful to remember that the kingdom of fungi constitutes a large and diverse group of organisms. The kingdom of fungi is divided into three sub-kingdoms which are further divided into phyla. Four of these are conjugating fungi, Ascomycota (sac fungi), Basidiomycota (club fungi) and Deuteromycotina (imperfect fungi). In this manner, all the mushrooms, molds, yeasts and mildew are all fungi. There are other phyla (Figure 7.1), but they are not as common.

7.1 NATIONAL AVERAGE

Over 200 fungi and some common contaminants are included in the fungi group and available for use in analysis. Remember that this list is growing, and as further fungi are sequenced, they can be added to the fungi group and the sample is re-analyzed. National average results show a consistent detection and identification of 75 fungi with an average range of 3–10 unique peptides per fungus, Figure 7.2. Several species of a particular fungus are included when available, since these different species can be found in different samples and sometimes in different locations. This number of unique peptides is actually low when compared with the average number of peptides for *Apis* (Chapter 3) and *Nosema* (Chapter 8). A sample from an average hive could be expected to fall with in this average for fungi.

Many fungal allergens are attributed to species of *Aspergillus fumigatus*, *Aspergillus flavus*, *Aspergillus nidulans*, *Alternaria alternata*, *Cladosporium cladosporoides*, *Ganoderma lucidum*, *Mucor mucedo*, *Fusarium solanii*, *Curvularia lunata*, *Neurospora sitophila*, *Scopulariopsis brevicaulis* and others. (Singh & Dahiya, 2008). The medical issues are numerous, but beyond the scope of this book. This list is included for information and also to indicate that these are common in the natural microflora.

The most common fungi species collected by honeybees are described in the following section in more detail.

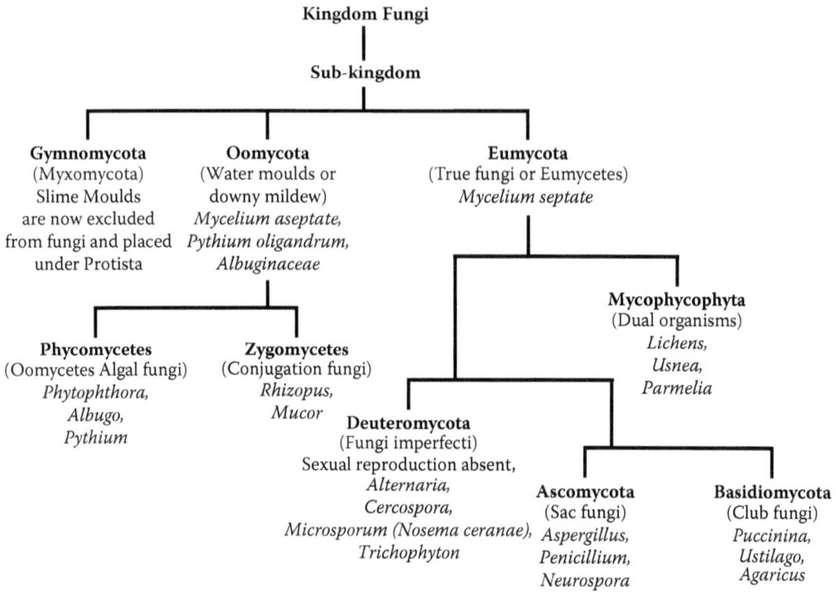

FIGURE 7.1 Classification of fungi.

7.1.1 *Alternaria alternata*

Alternaria alternata is a spore-forming fungi of the phylum Ascomycota.

Ascomycota (known as the ascomycetes) is the largest phylum of fungi, with over 64,000 species. The defining feature of this phylum is the "ascus", a microscopic sexual structure in which nonmotile spores are formed. Within the Ascomycota is the class Dothideomycetes, which is considered the largest and most diverse class of ascomycete fungi. It comprises 11 orders, 90 families, 1,300 genera and over 19,000 known species. Likewise, Pleosporales is the largest order in the fungal class Dothideomycetes, which contains 23 families, 332 genera and more than 4,700 species. The majority of species are saprobes on decaying plant material in fresh water, marine or terrestrial environments.

Alternaria alternata is one of these species known for causing leaf spot and other diseases on over 380 host species of plant. It is an opportunistic pathogen and is isolated from numerous kinds of organic materials in damp situations, including textiles, stored food, canvas, cardboard and paper, electric cables, polyurethane, jet fuel, sewage and effluents and is known for its saprophytic nature in decomposing soil and plants. *A. alternata* is ubiquitous.

The national average for *A. alternata* is 9.7 unique peptides. The California average is 7.9 and the Florida average is 3.4. The Idaho average is 11.0 unique peptides and Iowa has 15.2 unique peptides. The unique peptides for *A. alternata*

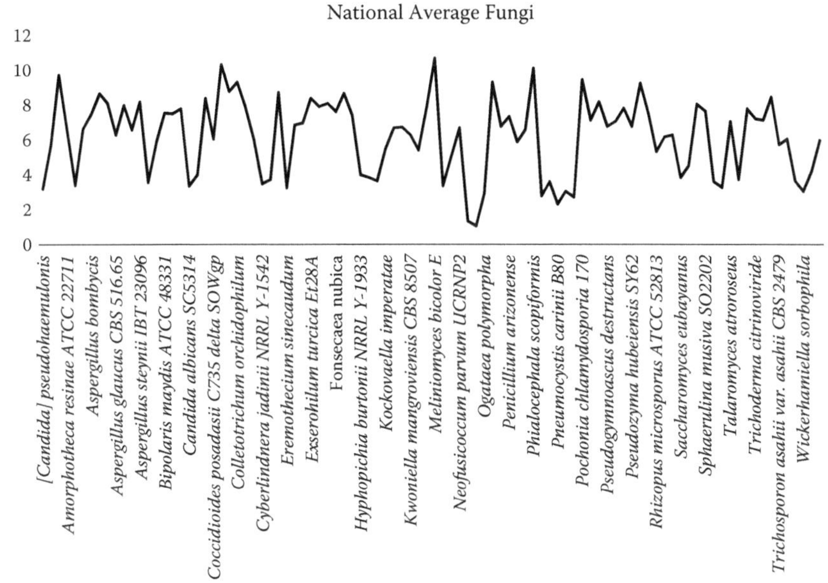

FIGURE 7.2 National unique peptide averages for fungi. Listing of all national fungi and their average of unique peptides is given in Appendix G.

are higher in the Montana region with an average of 19.0 (Figure 7.3). Examination of Figure 7.2, the national average for fungi, illustrates the average for all fungi.

7.1.2 *Aspergillus bombycis*

Aspergillus bombycis is a spore-forming fungi of the phylum *Ascomycota*.

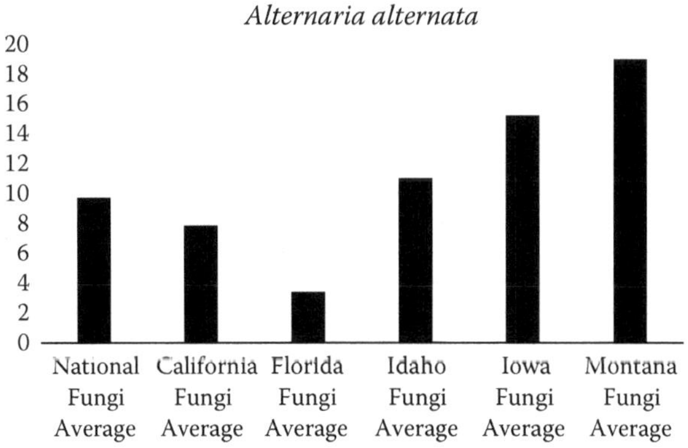

FIGURE 7.3 Detection and identification of *Alternaria alternata* using the MSP/ABOid method.

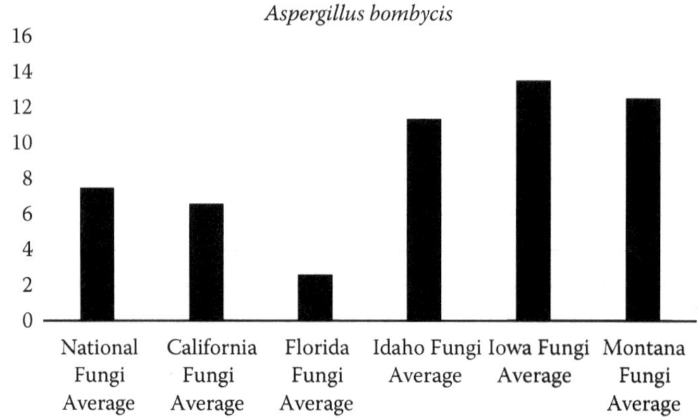

FIGURE 7.4 Detection and identification of *Aspergillus bombycis* using the MSP/ABOid method.

A. bombycis differs from *A. flavus* and *A. nomius* by a difference in their growth rates (Peterson, Ito, Horn, & Goto, 2001) and was isolated from silkworm-rearing houses in Japan and Indonesia. It is not clear that *A. bombycis* has been isolated from other sources, but *A. nomius* has been isolated from several sources including alkali bees (*Nomius* sp.), wheat and soil, a diseased pine sawfly and cotton field soil, including those in the United States.

The national average for *A. bombycis* is 7.5 unique peptides. The California average is 6.6 and the Florida average is 2.6. The Idaho average is 11.3 unique peptides and Iowa has 13.5 unique peptides. The unique peptides for *A. bombycis* are lower in the Montana region with an average of 12.5 (Figure 7.4). Examination of Figure 7.2, the national average for fungi, illustrates the average for all fungi.

7.1.3 *Aspergillus flavus, A. fumigatus* and *A. niger*

Aspergillus flavus, *A. fumigatus* and *A. niger* are spore-forming fungi of the phylum Ascomycota. They are included together because of their high average of unique peptides which are much higher than other fungi.

These three *Aspergillus* fungi are the most likely of the fungi to be found in the environment. They are known for the black mold of *A. niger* and the human respiratory difficulties of *A. fumigatus* and *A. flavus*. They are found throughout the environment, including soil, plant matter and household dust and have an essential role in carbon and nitrogen recycling

The national average for *A. flavus* is 97 unique peptides, *A. fumigatus* has 83 and *Aspergillus niger* 93. The California average for *A. flavus* is 72 unique peptides, *A. fumigatus* has 72 and *Aspergillus niger* 72. Florida average for *A. flavus* is 48 unique peptides, *A. fumigatus* has 42 and *A. niger* 47. The Idaho average for *A. flavus* is 96 unique peptides, *A. fumigatus* has 92 and *A. niger* 97. The unique peptides for *A. flavus* in the Iowa region are 167, *Aspergillus*

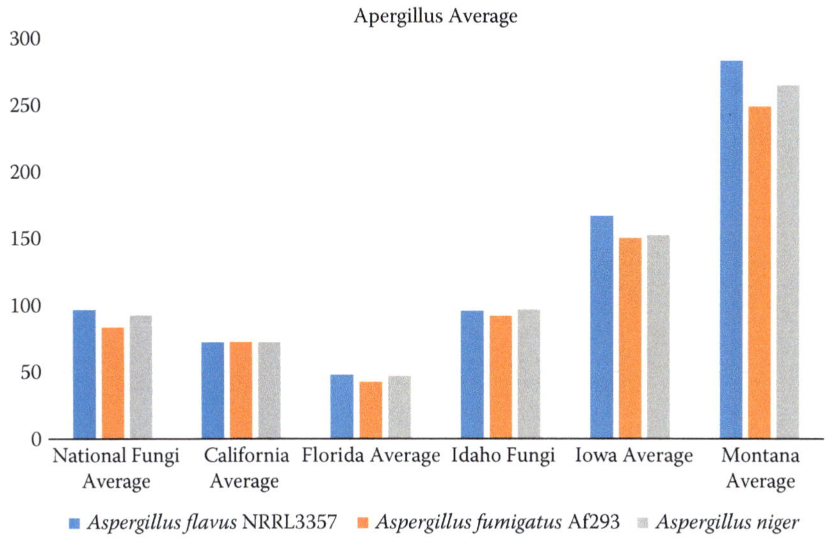

FIGURE 7.5 Detection and identification of *Aspergillus flavus*, *A. fumigatus* and *A. niger* using the MSP/ABOid method.

fumigatus has 150 and *Aspergillus niger* 152. Montana has the highest averages, with *Aspergillus flavus* averaging 283 unique peptides, *Aspergillus fumigatus* has 248 and *Aspergillus niger* 265 unique peptides (Figure 7.5).

7.1.4 *Aspergillus oryzae* RIB40

Aspergillus oryzae is a spore-forming fungus of the phylum Ascomycota.

A. oryzae is a domesticated species and are most commonly found in northern regions, specifically in East Asia, where it is used in solid-state cultivation, soybean fermentation and for sake brewing. Although associated with Japan and China, *Aspergillus oryzae* can be found anywhere. Although the *Aspergillus* genus is extremely common, *A. oryzae* is considered to be rare due to its domestication for use in fermentation and in the food industry.

The high average of unique peptides for *A. oryzae* found on the honeybees would give cause for further investigation or at least add to the history of where this fungus can be found. It is found in all our regions and in high enough numbers to be considered a normal occurring fungus among the honeybee samples.

The national average for *A. oryzae* is 6.6 unique peptides. The California average is 5.1 and the Florida average is 3.1. The Idaho average is 6.7 unique peptides and Iowa has 12.2 unique peptides. The unique peptides for *A. oryzae* are higher in the Montana region with an average of 14.0 (Figure 7.6). Examination of Figure 7.2, the national average for fungi, illustrates the average for all fungi.

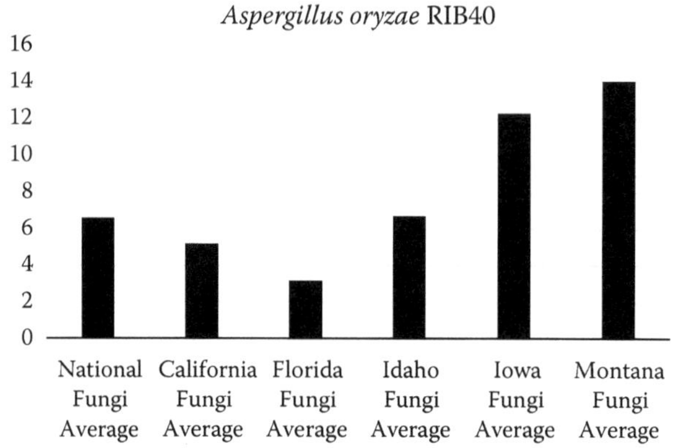

FIGURE 7.6 Detection and identification of *Aspergillus oryzae* using the MSP/ABOid method.

7.1.5 *Aspergillus steynii* IBT 23096

Aspergillus steynii is a species of fungus in the phylum Ascomycota.

A. steynii has been found in Argentina, Australia, China, India, Panama, Spain and Sri Lanka on chili, coffee beans, grapes, paprika, rice, soy beans and in soil.

A. steynii is also found in five regions of the United States and has a national average of 8.2 unique peptides. The California average is 8.3 and the Florida average is 3.6. The Idaho average is 10.7 unique peptides and Iowa has 14.1 unique peptides. The unique peptides for *A. steynii* are lower in the Montana region with an average of 13.3 (Figure 7.7). Examination of Figure 7.2, the national average for fungi, illustrates the average for all fungi. The Iowa and Montana regions show some of the highest unique peptides for *A. steynii.*

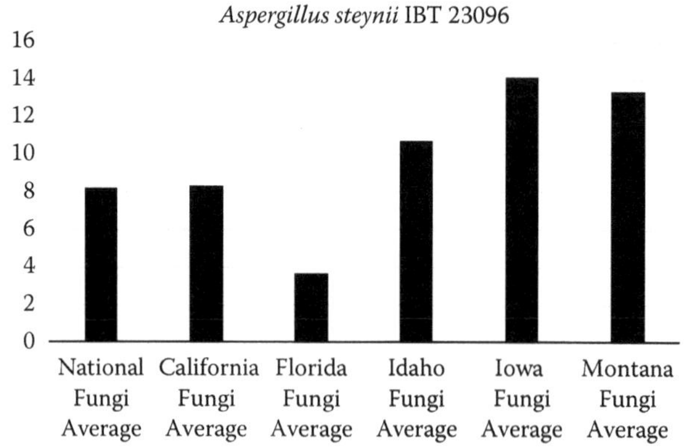

FIGURE 7.7 Detection and identification of *Aspergillus steynii* using the MSP/ABOid method.

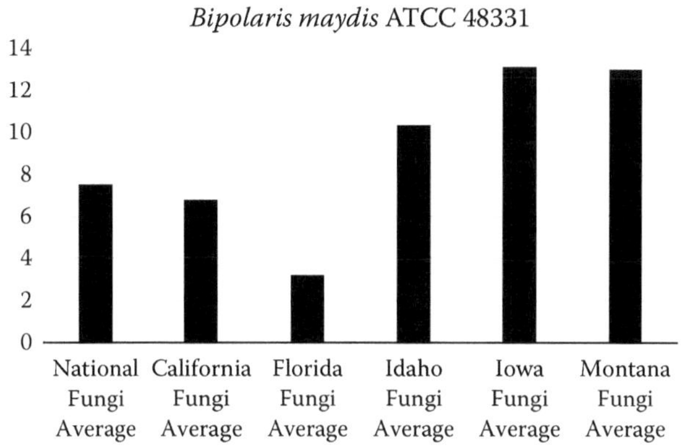

FIGURE 7.8 Detection and identification of *Bipolaris maydis* using the MSP/ABOid method.

7.1.6 *Bipolaris maydis* ATCC 48331

Bipolaris maydis is a species of fungus in the phylum Ascomycota.

B. maydis is also known as *Drechslera maydis* or *Cochliobolus heterostrophus*, causes Southern corn leaf blight and stalk rot. Since it can overwinter by surviving on plant debris, it is persistent in the environment as seen in Figure 7.8.

The national average for *Bipolaris maydis* is 7.5 unique peptides. The California average is 6.8 and the Florida average is 3.2. The Idaho average is 10.3 unique peptides and Iowa has 13.1 unique peptides. The unique peptides for *bipolaris maydis* are lower in the Montana region with an average of 13.0 (Figure 7.8). Examination of Figure 7.2, the national average for fungi, illustrates the average for all fungi.

7.1.7 *Bipolaris sorokiniana* ND90Pr

Bipolaris sorokiniana is a species of fungus in the phylum Ascomycota.

B. sorokiniana like other Ascomycetes is widely distributed and is the causal agent for many plant diseases of cereal, common root rot, and leaf and stem infections most notably on wheat and barley crops.

The national average for *B. sorokiniana* is 7.5 unique peptides. The California average is 6.9 and the Florida average is 3.2. The Idaho average is 10.3 unique peptides and Iowa has 13.0 unique peptides. The unique peptides for *B. sorokiniana* are lower in the Montana region with an average of 11.5 (Figure 7.9). Examination of Figure 7.2, the national average for fungi, illustrates the average for all fungi.

7.1.8 *Botrytis cinerea* B05.10

Botrytis cinerea is a species of fungus in the phylum Ascomycota.

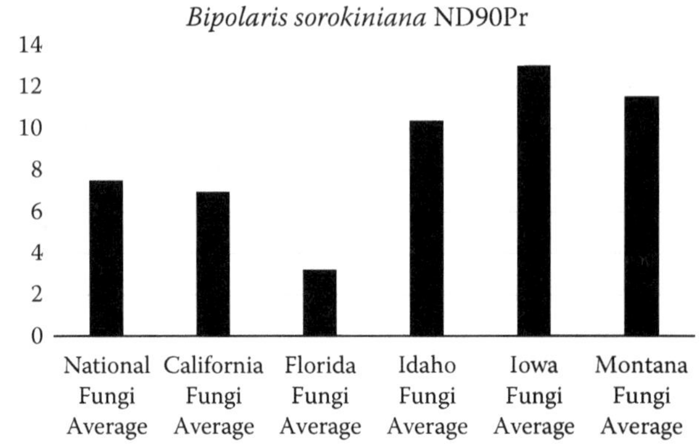

FIGURE 7.9 Detection and identification of *Bipolaris sorokiniana* using the MSP/ABOid method.

B. cinerea produces a highly resistant sclerotia stage as a survival structures and can overwinter as either sclerotia or intact mycelia which germinate in the spring to produce conidiophores. The conidia are dispersed by the wind and to cause new infections.

B. cinerea is known to infect a large number of plants including protein crops, fiber crops, oil crops and horticultural crops with a gray mold. Horticultural crops include chickpeas, lettuce, broccoli and beans; also, grapes, strawberries and raspberries.

The national average for *B. cinerea* is 7.8 unique peptides. The California average is 6.4 and the Florida average is 3.2. The Idaho average is 10.0 unique peptides and Iowa has 13.7 unique peptides. The unique peptides for *B. cinerea* are lower in the Montana region with an average of 11.8 (Figure 7.10).

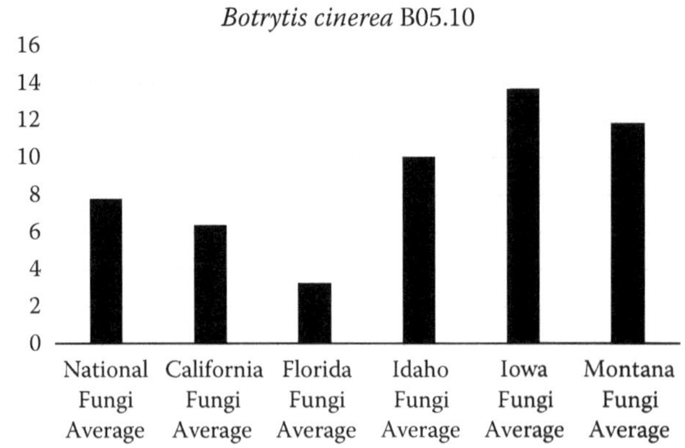

FIGURE 7.10 Detection and identification of *Botrytis cinerea* using the MSP/ABOid method.

Examination of Figure 7.2, the national average for fungi, illustrates the average for all fungi.

7.1.9 *Cercospora beticola*

Cercospora beticola is a species of fungus in the phylum Ascomycota.

C. beticola is the cause of Cercospora leaf spot disease in sugar beets, spinach and swiss chard and kills the plants by preventing root formation.

Like other ascomycetes, *C. beticola* produces a sclerotia-like overwintering structures (stromata) that contain conidia. Under favorable wet conditions they germinate and start the disease cycle by infecting the host plants and then produce sclerotia again at the end of the growing season.

The national average for *C. beticola* is 8.4 unique peptides. The California average is 7.2 and the Florida average is 3.2. The Idaho average is 13.7 unique peptides and Iowa has 13.5 unique peptides. The unique peptides for *C. beticola* are higher in the Montana region with an average of 15 (Figure 7.11). Examination of Figure 7.2, the national average for fungi, illustrates the average for all fungi.

7.1.10 *Colletotrichum gloeosporioides* Nara gc5

Colletotrichum gloeosporioides is a species of fungus in the phylum Ascomycota.

C. gloeosporioides has a broad host range and is responsible for fruit-rotting diseases on hundreds of economically important crop plants including cereals, grasses, legumes, fruits, vegetables, trees and other crops. *C. gloeosporioides* is distributed worldwide.

The national average for *C. gloeosporioides* is 10.3 unique peptides. The California average is 7.9 and the Florida average is 4.6. The Idaho average is 10.7 unique peptides and Iowa has 16.2 unique peptides. The unique peptides for

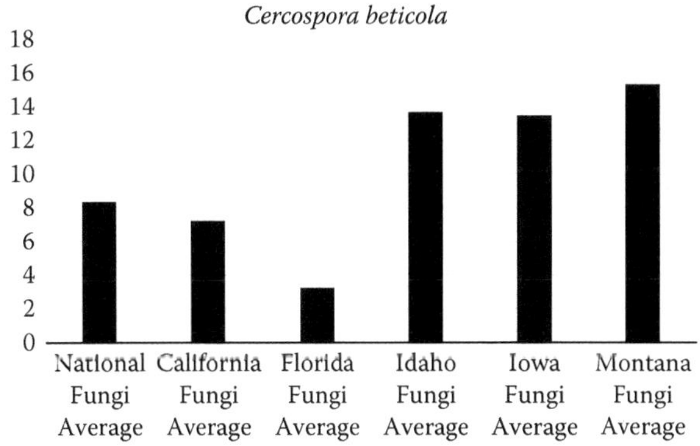

FIGURE 7.11 Detection and identification of *Cercospora beticola* using the MSP/ABOid method.

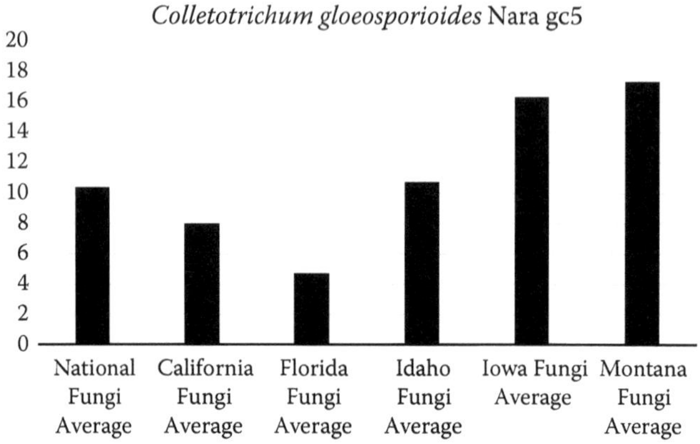

FIGURE 7.12 Detection and identification of *Colletotrichum gloeosporioides* using the MSP/ABOid method.

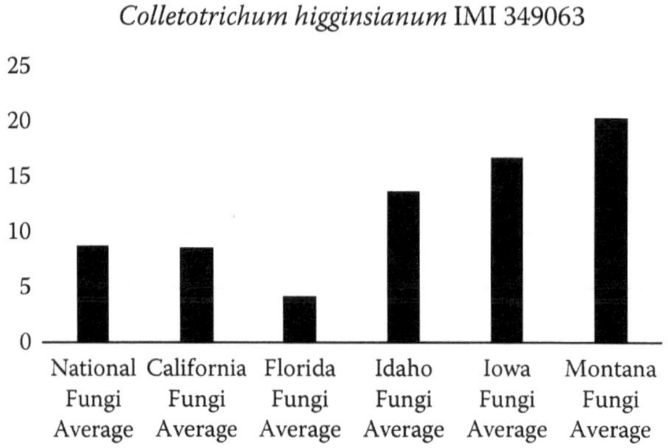

FIGURE 7.13 Detection and identification of *Colletotrichum higginsianum* using the MSP/ABOid method.

C. gloeosporioides are higher in the Montana region with an average of 17.3 (Figure 7.12). Examination of Figure 7.2, the national average for fungi, illustrates the average for all fungi.

7.1.11 *Colletotrichum higginsianum* IMI 349063

Colletotrichum higginsianum is a species of fungus in the phylum Ascomycota.

Colletotrichum is a large ascomycete genus comprising more than 190 species many of which cause diseases on a large range of agricultural and horticultural crops worldwide.

C. higginsianum causes anthracnose (disease name of several common fungal diseases) on a wide range of cruciferous plants, such as species of *Brassica* (cabbage, turnip, Brussels sprout, and mustard) and *Raphanus* (radish).

The national average for *C. higginsianum* is 8.8 unique peptides. The California average is 8.6 and the Florida average is 4.1. The Idaho average is 13.7 unique peptides and Iowa has 16.7 unique peptides. The unique peptides for *C. higginsianum* are higher in the Montana region with an average of 20.3 (Figure 7.13). Examination of Figure 7.2, the national average for fungi, illustrates the average for all fungi.

7.1.12 *COLLETOTRICHUM ORCHIDOPHILUM*

Colletotrichum orchidophilum is a species of fungus in the phylum Ascomycota.

Colletotrichum is a large ascomycete genus comprising more than 190 species many of which cause diseases on a large range of agricultural and horticultural crops worldwide.

C. orchidophilum causes diseases on several species of the family Orchidaceae (flowering plants).

The national average for *C. orchidophilum* is 9.3 unique peptides. The California average is 7.4 and the Florida average is 3.2. The Idaho average is 10.7 unique peptides and Iowa has 14.5 unique peptides. The unique peptides for *C. orchidophilum* are higher in the Montana region with an average of 15.3 (Figure 7.14). Examination of Figure 7.2, the national average for fungi, illustrates the average for all fungi.

7.1.13 *DIPLODIA CORTICOLA*

Diplodia corticola is the asexual stage of *Botryosphaeria corticola*, a species of fungus in the phylum Ascomycota.

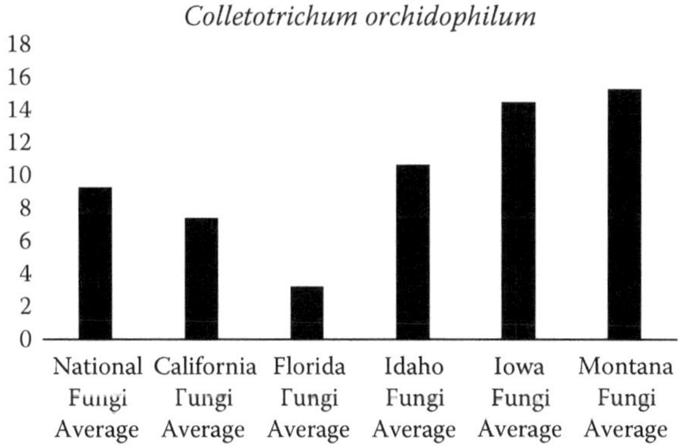

FIGURE 7.14 Detection and identification of *Colletotrichum orchidophilum* using the MSP/ABOid method.

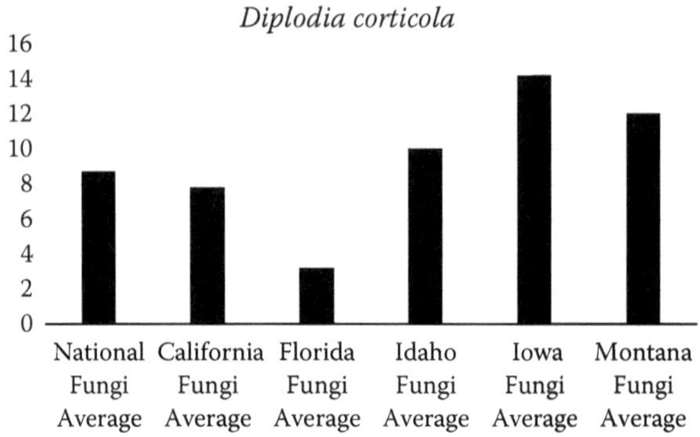

FIGURE 7.15 Detection and identification of *Diplodia corticola* using the MSP/ABOid method.

D. corticola is the causative agent of "Bot canker" of oak and occurs on stems, branches and twigs of oak trees in Europe and North America.

The national average for *D. corticola* is 8.7 unique peptides. The California average is 7.8 and the Florida average is 3.2. The Idaho average is 10.0 unique peptides and Iowa has 14.2 unique peptides. The unique peptides for *D. corticola* are lower in the Montana region with an average of 12.0 (Figure 7.15). Examination of Figure 7.2, the national average for fungi, illustrates the average for all fungi.

7.1.14 *Exserohilum turcica* Et28A

Exserohilum turcica is a species of fungus in the phylum Ascomycota and the family Dipodascaceae.

The Dipodascaceae are a family of yeasts in the order Saccharomycetales. Species in the family have a widespread distribution and are found in decaying plant tissue, or as spoilage organisms in the food industry.

Exserohilum turcicum (type species) causes Northern corn leaf blight (NCLB), which is a foliar disease of corn (maize). *E. turcicum* is the anamorph (asexual reproductive stage) of the ascomycete *Setosphaeria turcica*. The disease is characterized by cigar-shaped lesions and can cause significant yield loss in susceptible corn hybrids.

The national average for *E. turcica* is 8.4 unique peptides. The California average is 7.8 and the Florida average is 3.4. The Idaho average is 12.7 unique peptides and Iowa has 14.8 unique peptides. The unique peptides for *E. turcica* are lower in the Montana region with an average of 10.5 (Figure 7.16). Examination of Figure 7.2, the national average for fungi, illustrates the average for all fungi.

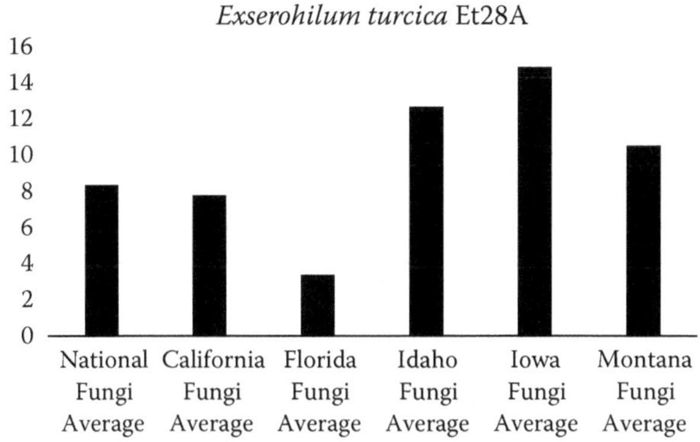

FIGURE 7.16 Detection and identification of *Exserohilum turcica* using the MSP/ABOid method.

7.1.15 *Fonsecaea erecta*

Fonsecaea erecta is a species of fungus in the phylum Ascomycota.

F. erecta, along with other ascomycetes in this family and the type species *Fonsecaea pedrosoi*, occurs in soil and on plants and trees where it frequently grows as a saprotroph.

Although *F. erecta* is associated with tropical and sub-tropical regions, Figure 7.17 shows a healthy distribution of *F. erecta* within the five regions of the United States. The national average for *F. erecta* is 7.9 unique peptides. The California average is 8.1 and the Florida average is 2.8. The Idaho average is 9.7 unique peptides and Iowa has 13.8 unique peptides. The unique peptides for *F. erecta* are lower in the Montana region with an average of 13.3 (Figure 7.17). Examination of Figure 7.2, the national average for fungi, illustrates the average for all fungi.

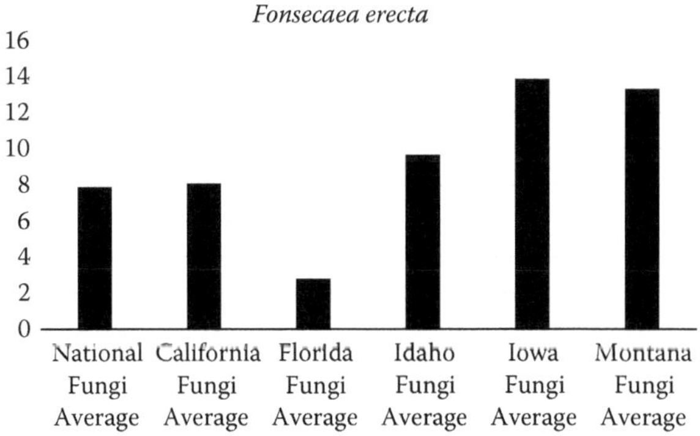

FIGURE 7.17 Detection and identification of *Fonsecaea erecta* using the MSP/ABOid method.

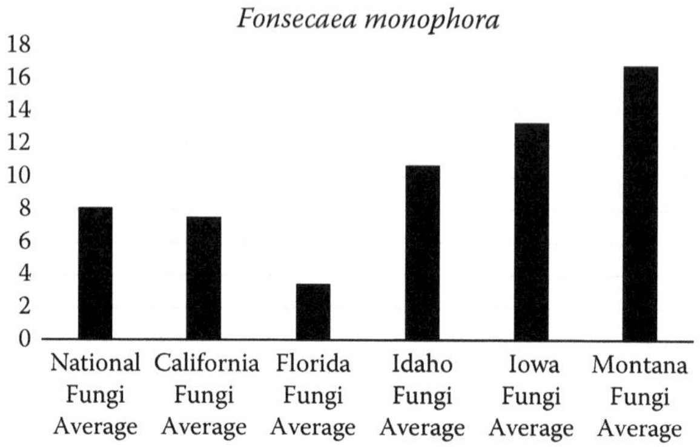

FIGURE 7.18 Detection and identification of *Fonsecaea monophora* using the MSP/ABOid method.

7.1.16 *Fonsecaea monophora*

Fonsecaea monophora is a species of fungus in the phylum Ascomycota.

Like other *Fonsecaea* and other ascomycetes in this family, *F. monophora* occurs in soil and on plants and trees where it frequently grows as a saprotroph. It is widely distributed and as can be seen in Figure 7.18 is present in all the regions.

The national average for *F. monophora* is 7.1 unique peptides. The California average is 7.5 and the Florida average is 3.4. The Idaho average is 10.7 unique peptides and Iowa has 13.3 unique peptides. The unique peptides for *F. monophora* are higher in the Montana region with an average of 16.8 (Figure 7.18). Examination of Figure 7.2, the national average for fungi, illustrates the average for all fungi.

7.1.17 *Fonsecaea nubica*

Fonsecaea nubica is a species of fungus in the phylum Ascomycota.

F. nubica is distributed worldwide and occurs in soil and on plants and trees where it frequently grows as a saprotroph.

Figure 7.19 illustrates the distribution within the United States and over five regions. The national average for *F. nubica* is 7.6 unique peptides. The California average is 6.9 and the Florida average is 3.2. The Idaho average is 9.3 unique peptides and Iowa has 13.5 unique peptides. The unique peptides for *F. nubica* are higher in the Montana region with an average of 15.5 (Figure 7.19). Examination of Figure 7.2, the national average for fungi, illustrates the average for all fungi.

7.1.18 *Fusarium fujikuroi* IMI 58289

Fusarium fujikuroi, also known as *Gibberella fujikuroi*, is a species of fungus in the phylum Ascomycota.

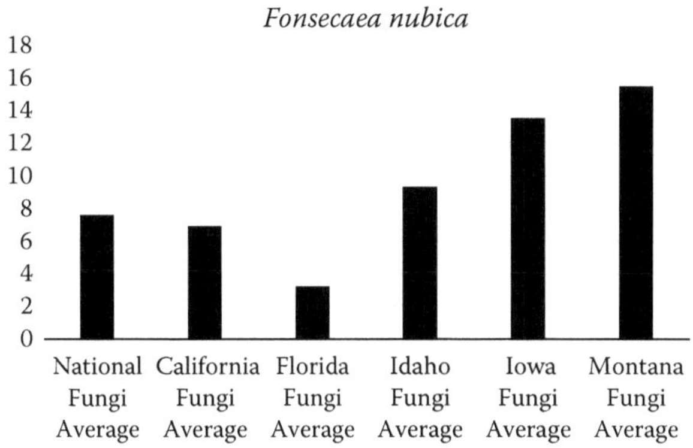

FIGURE 7.19 Detection and identification of *Fonsecaea nubica* using the MSP/ABOid method.

F. fujikuroi is ubiquitous in nature, where it is extensively distributed in soil, plants and various organic substances where most of it is a harmless saprobe.

F. fujikuroi produces the plant hormone gibberellin that causes excessive growth and poor yield, particularly in rice plants; but barley, millet, sugarcane and maize are susceptible (Hsuan, Salleh, & Zakaria, 2011). An interesting alternate name is "foolish seedling disease" because of the disparity of the growth in the infected seedling.

Although associated with Asian countries such as India, Thailand and Japan, it is obviously distributed within the United States (Figure 7.20) and maybe elsewhere. The collection of this ascomycete by the honeybee shows a distribution among all regions.

The national average for *F. fujikuroi* is 8.6 unique peptides. The California average is 7.6 and the Florida average is 4.3. The Idaho average is 11.7 unique

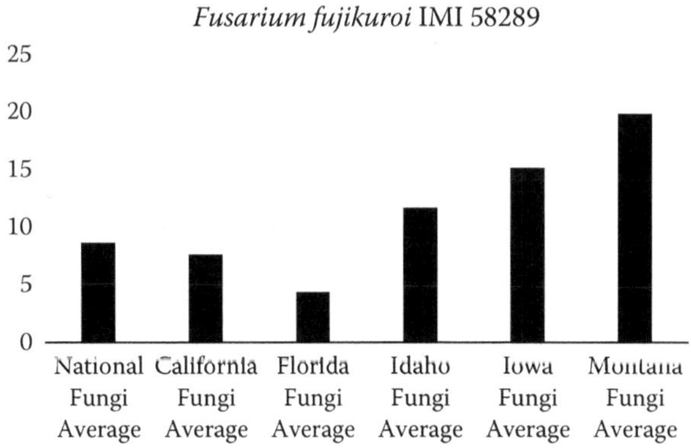

FIGURE 7.20 Detection and identification of *Fusarium fujikuroi* using the MSP/ABOid method.

peptides and Iowa has 15.1 unique peptides. The unique peptides for *F. fujikuroi* are higher in the Montana region with an average of 19.8 (Figure 7.20). Examination of Figure 7.2, the national average for fungi, illustrates the average for all fungi.

7.1.19 *Meliniomyces bicolor* E

Meliniomyces bicolor is a species of fungus in the phylum Ascomycota.

Meliniomyces has been isolated from the roots of the Orchidaceae (orchids), Pinaceae (pine family that includes cedars, firs, hemlocks, larches, pines and spruces), Betulaceae (birches, alders, hazels, hornbeams, hazel-hornbeam and hop-hornbeams) and Salicaceae (willows, poplar, aspen and cottonwoods). *M. bicolor* is closely associated with many Northern temperate forest trees, such as pine, spruce and birch.

The national average for *M. bicolor* is 10.7 unique peptides. The California average is 8.6 and the Florida average is 4.2. The Idaho average is 13.0 unique peptides and Iowa has 17.4 unique peptides. The unique peptides for *M. bicolor* are lower in the Montana region with an average of 16.3 (Figure 7.21). Examination of Figure 7.2, the national average for fungi, illustrates the average for all fungi.

7.1.20 *Paraphaeosphaeria sporulosa*

Paraphaeosphaeria sporulosa is a species of fungus in the phylum Ascomycota. Anamorph (asexual stage) forms are found in the genus *Paraconiothyrium.*

Paraphaeosphaeria is an endophytic fungus known to produce anti-bacterial metabolites. It has 23 species found in Europe and North America.

P. sporulosa is widely distributed among the five regions within the United States (Figure 8.1). The national average for *P. sporulosa* is 9.3 unique peptides.

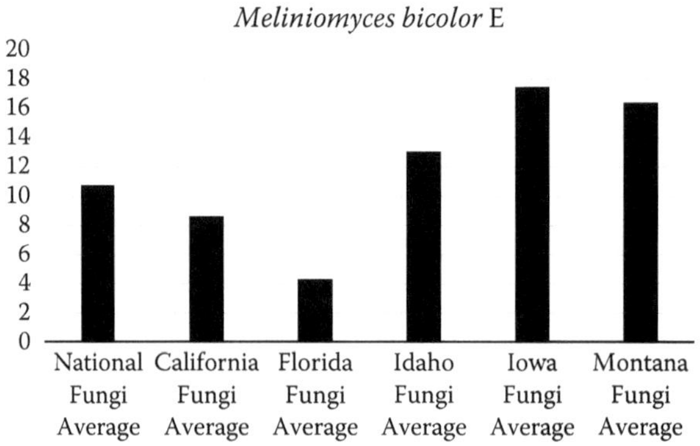

FIGURE 7.21 Detection and identification of *Meliniomyces bicolor* using the MSP/ABOid method.

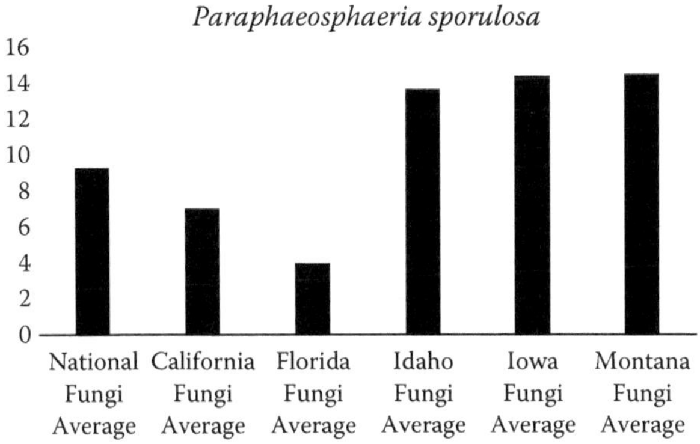

FIGURE 7.22 Detection and identification of *Paraphaeosphaeria sporulosa* using the MSP/ABOid method.

The California average is 7.0 and the Florida average is 4.0. The Idaho average is 13.7 unique peptides and Iowa has 14.4 unique peptides. The unique peptides for *P. sporulosa* are higher in the Montana region with an average of 14.5 (Figure 7.22). Examination of Figure 7.2, the national average for fungi, illustrates the average for all fungi.

7.1.21 *Penicillium arizonense*

Penicillium arizonense is a species of fungus in the phylum Ascomycota.

P. arizonense is a species with a high chemical diversity in secreted biomass-degrading enzymes, some of which are involved in carbohydrate metabolism, in particular hemicellulases.

The national average for *P. arizonense* is 7.3 unique peptides. The California average is 6.3 and the Florida average is 3.0. The Idaho average is 13.7 unique peptides and Iowa has 13.1 unique peptides. The unique peptides for *P. arizonense* are lower in the Montana region with an average of 11.8 (Figure 7.23). Examination of Figure 7.2, the national average for fungi, illustrates the average for all fungi.

7.1.22 *Phialocephala scopiformis*

Phialocephala scopiformis is a species of fungus in the phylum Ascomycota.

P. scopiformis is a conifer needle endophyte and produces a potent anti-insectan metabolite rugulosin. As a result, infected trees are more tolerant to spruce budworm, a serious insect pest of forests in northeastern United States and eastern Canada.

As with most ascomycetes, *P. scopiformis* is widely distributed within the United States (Figure 7.24). The national average for *P. scopiformis* is 10.1 unique peptides. The California average is 8.9 and the Florida average is 4.1. The Idaho average is 15.0 unique peptides and Iowa has 17.3 unique peptides. The

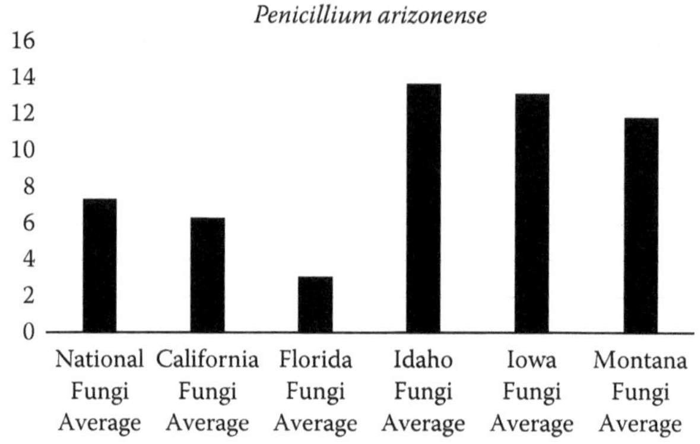

FIGURE 7.23 Detection and identification of *Penicillium arizonense* using the MSP/ABOid method.

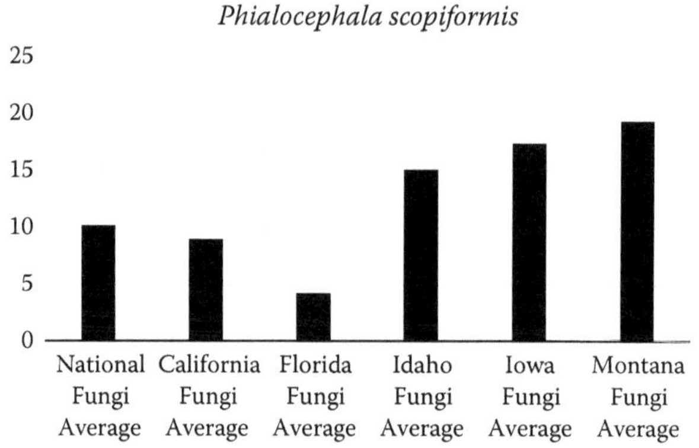

FIGURE 7.24 Detection and identification of *Phialocephala scopiformis* using the MSP/ABOid method.

unique peptides for *P. scopiformis* are higher in the Montana region with an average of 19.3 (Figure 7.24). Examination of Figure 7.2, the national average for fungi, illustrates the average for all fungi.

7.1.23 *Pochonia chlamydosporia* 170

Pochonia chlamydosporia is a species of fungus in the phylum Ascomycota.

P. chlamydosporia colonizes tomato roots and is a fungal parasite of root-knot nematodes eggs that can endophytically colonize the roots of several cultivated plant species.

P. chlamydosporia is widely dispersed within the United States as illustrated in Figure 7.25. The national average for *P. chlamydosporia* is 9.4 unique

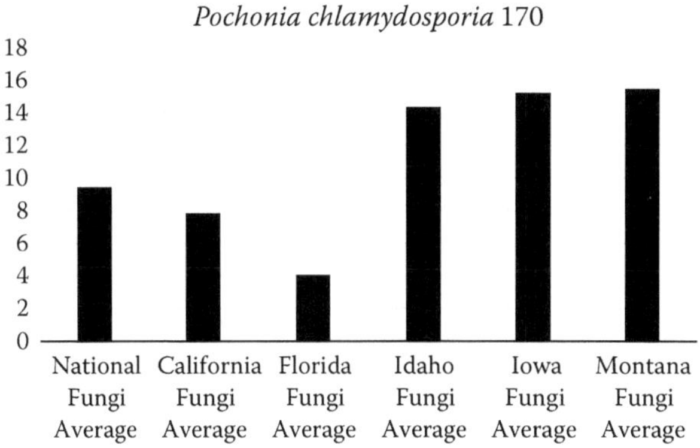

FIGURE 7.25 Detection and identification of *Pochonia chlamydosporia* using the MSP/ABOid method.

peptides. The California average is 7.9 and the Florida average is 4.0. The Idaho average is 14.3 unique peptides and Iowa has 15.2 unique peptides. The unique peptides for *P. chlamydosporia* are higher in the Montana region with an average of 15.5 (Figure 7.25). Examination of Figure 7.2, the national average for fungi, illustrates the average for all fungi.

7.1.24 *Pseudocercospora fijiensis* CIRAD86

Pseudocercospora fijiensis is a species of fungus in the phylum Ascomycota.

P. fijiensis is known to cause black leaf streak on banana leaves and has spread rapidly worldwide. It has been reported in Florida.

The spread of this fungus has continued, and it would appear that it has found other hosts other than banana leaves as shown in Figure 7.26, where it is seen at different levels in all five regions, including the northern regions of Idaho and Montana. It was collected by the honeybees from some repository.

The national average for *P. fijiensis* is 8.2 unique peptides. The California average is 7.1 and the Florida average is 14.0. The Idaho average is 11.3 unique peptides and Iowa has 13.5 unique peptides. The unique peptides for *P. fijiensis* are lower in the Montana region with an average of 13.5 (Figure 7.26). Examination of Figure 7.2, the national average for fungi, illustrates the average for all fungi.

7.1.25 *Purpureocillium lilacinum*

Purpureocillium lilacinum is purple conidia-forming fungus in the phylum Ascomycota.

P. lilacinum has been isolated from several habitats, including cultivated and uncultivated soils, forests, grassland, deserts, estuarine sediments, sewage sludge and insects. It has also been found in nematode eggs and the rhizosphere of many crops.

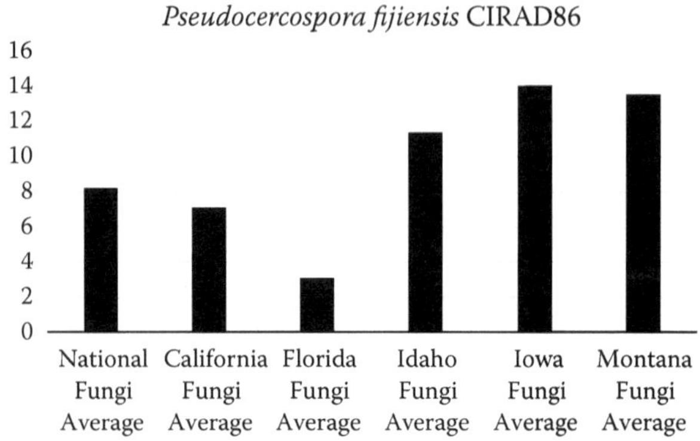

FIGURE 7.26 Detection and identification of *Pseudocercospora fijiensis* using the MSP/ABOid method.

The national average for *P. lilacinum* is 9.2 unique peptides. The California average is 7.8 and the Florida average is 2.8. The Idaho average is 13.0 unique peptides and Iowa has 14.1 unique peptides. The unique peptides for *P. lilacinum* are higher in the Montana region with an average of 17.8 (Figure 7.27). Examination of Figure 7.2, the national average for fungi, illustrates the average for all fungi.

7.1.26 *Ramularia collo-cygni*

Ramularia collo-cygni is a genus of fungi that are anamorphs of the genus *Mycosphaerella* in the phylum Ascomycota.

R. collo-cygni is a plant pathogen, and hosts include the genus *Narcissus* (daffodil, narcissus and jonquil; Kirk, Cannon, Minter, & Stalpers, 2008).

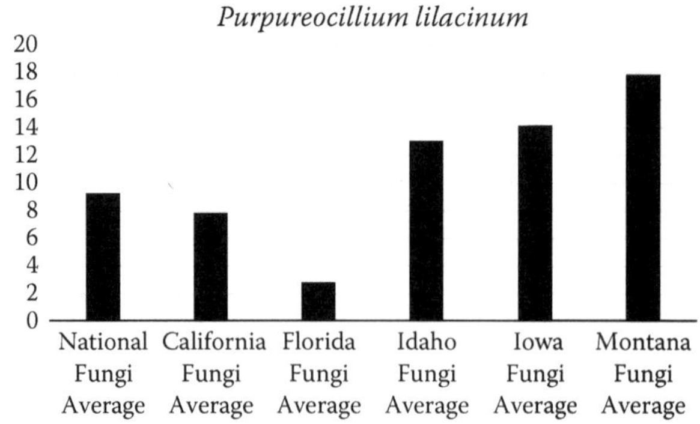

FIGURE 7.27 Detection and identification of *Purpureocillium lilacinum* using the MSP/ABOid method.

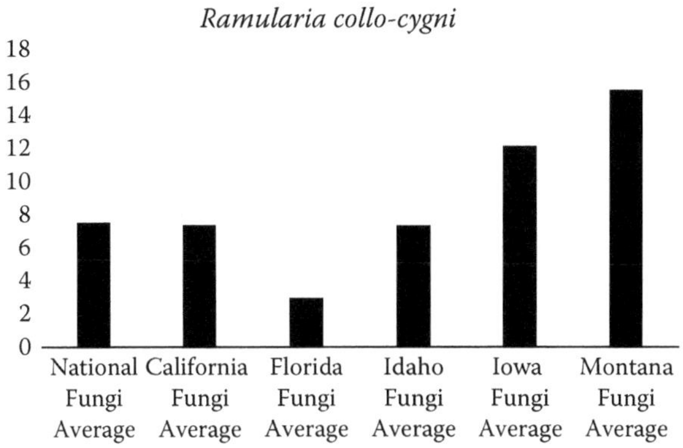

FIGURE 7.28 Detection and identification of *Ramularia collo-cygni* using the MSP/ABOid method.

R. collo-cygni is found in all five regions across the United States (Figure 7.28) with the greatest numbers in Iowa and Montana. The national average for *R. collo-cygni* is 7.5 unique peptides. The California average is 7.4 and the Florida average is 3.0. The Idaho average is 7.3 unique peptides and Iowa has 12.1 unique peptides. The unique peptides for *R. collo-cygni* are higher in the Montana region with an average of 15.5 (Figure 7.28). Examination of Figure 7.2, the national average for fungi, illustrates the average for all fungi.

7.1.27 *Rhizopus microsporus* ATCC 52813

Rhizopus microsporus is a fungus of the phylum Mucoromycota.

R. microsporus is a widely distributed plant pathogen associated with infecting maize (Rhizopus ear rot), sunflower (head rot) and rice (rice seedling blight). The fungus has other activities such as in the preparation of soy fermentation and in the production of tempeh and sufu. *R. microsporus* is commonly found in soil, plant debris and many diverse environments.

R. microsporus contains the bacterial endosymbiont *Burkholderia rhizoxinica* which produces rhizoxin. The fungus and the bacteria may be detected together.

The national average for *R. microsporus* is 5.3 unique peptides. The California average is 4.6 and the Florida average is 1.9. The Idaho average is 6.0 unique peptides and Iowa has 9.1 unique peptides. The unique peptides for *R. microsporus* are higher in the Montana region with an average of 13.5 (Figure 7.29). Examination of Figure 7.2, the national average for fungi, illustrates the average for all fungi.

7.1.28 *Talaromyces atroroseus*

Talaromyces atroroseus is a species of fungus in the phylum Ascomycota.

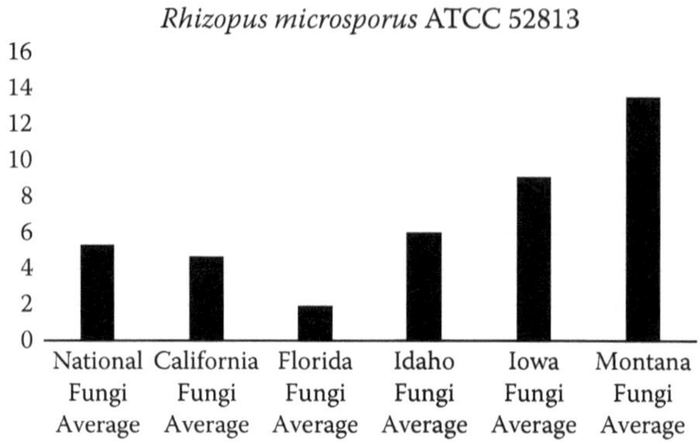

FIGURE 7.29 Detection and identification of *Rhizopus microsporus* using the MSP/ABOid method.

T. atroroseus is found in soil and fruit. It is known for the ability to produce a stable red pigment and has no known toxins.

Although associated with rural areas in most Southeast Asian countries, including Thailand, Vietnam, Laos and southern China, it is also located across the five regions of the United States as shown in Figure 7.30.

The national average for *T. atroroseus* is 7.0 unique peptides. The California average is 5.8 and the Florida average is 3.8. The Idaho average is 9.0 unique peptides and Iowa has 10.8 unique peptides. The unique peptides for *T. atroroseus* are higher in the Montana region with an average of 14.3 (Figure 7.30). Examination of Figure 7.2, the National average for fungi, illustrates the average for all fungi.

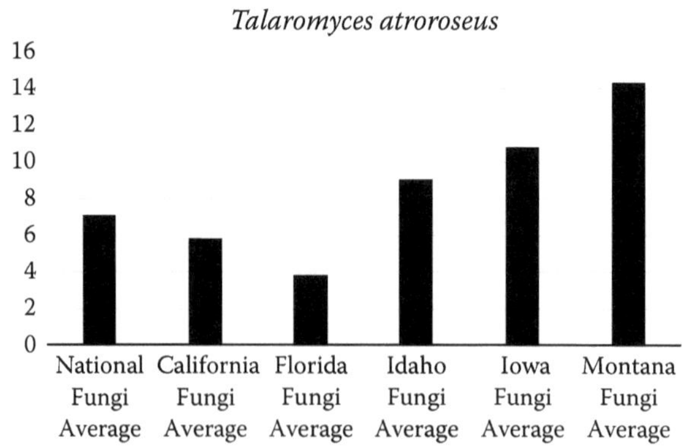

FIGURE 7.30 Detection and identification of *Talaromyces atroroseus* using the MSP/ABOid method.

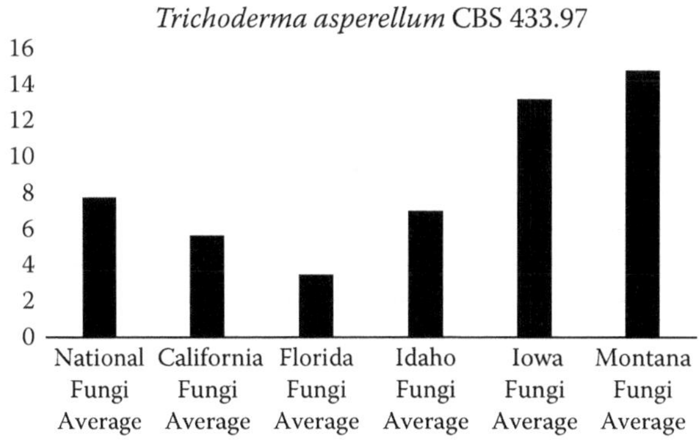

FIGURE 7.31 Detection and identification of *Trichoderma asperellum* using the MSP/ABOid method.

7.1.29 *Trichoderma asperellum* CBS 433.97

Trichoderma asperellum is a fungus of the phylum Ascomycota.

T. asperellum has been used commercially as a biopesticide for plant disease control. It is present in soil and is known as the most prevalent culturable fungi in the soil. They are symbionts and have the ability to form endophytic relationships with several plant species, which makes *T. asperellum* suitable for wide distribution as seen in Figure 7.31.

The national average for *T. asperellum* is 7.8 unique peptides. The California average is 5.6 and the Florida average is 3.5. The Idaho average is 7.0 unique peptides and Iowa has 13.2 unique peptides. The unique peptides for *T. asperellum* are higher in the Montana region with an average of 14.8 (Figure 7.31). Examination of Figure 7.2, the national average for fungi, illustrates the average for all fungi.

7.1.30 *Trichoderma gamsii*

Trichoderma gamsii is a fungus of the phylum Ascomycota.

T. gamsii, like other species of this genus, is an endophytic fungus associated with mountain ecosystems, although it appears to be equally associated with other types of ecology as illustrated in Figure 7.32.

The national average for *T. gamsii* is 7.1 unique peptides. The California average is 4.2 and the Florida average is 3.6. The Idaho average is 8.3 unique peptides and Iowa has 11.5 unique peptides. The unique peptides for *T. gamsii* are higher in the Montana region with an average of 14.5 (Figure 7.32). Examination of Figure 7.2, the national average for fungi, illustrates the average for all fungi.

7.1.31 *Trichoderma harzianum* CBS 226.95

Trichoderma harzianum is a fungus of the phylum Ascomycota.

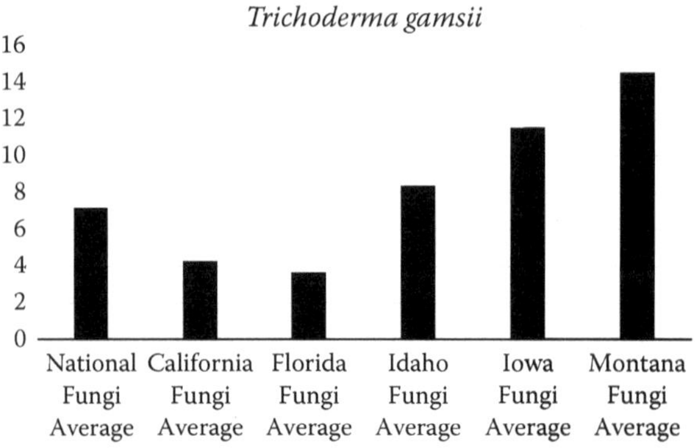

FIGURE 7.32 Detection and identification of *Trichoderma gamsii* using the MSP/ABOid method.

T. harzianum is known for its use as a fungicide, both in seed and soil treatments. *Trichoderma* spp. are present in nearly all soils and many other diverse habitats. *T. harzianum* can be expected to be widely discriminated and that is illustrated in Figure 7.33.

The national average for *T. harzianum* is 8.4 unique peptides. The California average is 7.1 and the Florida average is 3.5. The Idaho average is 10.7 unique peptides and Iowa has 14.8 unique peptides. The unique peptides for *T. harzianum* are higher in the Montana region with an average of 16.3 (Figure 7.33). Examination of Figure 7.2, the national average for fungi, illustrates the average for all fungi.

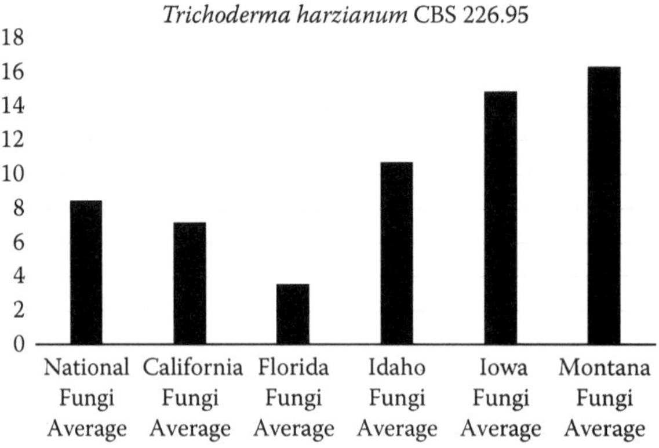

FIGURE 7.33 Detection and identification of *Trichoderma harzianum* using the MSP/ABOid method.

7.2 DISCUSSION ON NATIONAL AVERAGE FOR FUNGI

One particular species of fungi does not appear to have an advantage over another. For example, *Aspergillus sp.* is a filamentous, cosmopolitan and ubiquitous fungus found in nature. There are over 185 species of *Aspergillus* of which many are frequently seen as the dark mold on all types of media. Many people first encounter it as the frequently seen black bread mold. In the case of honeybees, it would not be surprising to see *Aspergillus* occurring in honeybee samples first. This appears, however, not to be the case when examining hundreds of honeybee samples and comparing the averages of different species and strains of fungi. *Coccidioides posadasii* appears in similar numbers and just as frequently as *Aspergillus* among samples. *Coccidiodies* is a pathogenic fungus that resides in semiarid soil typically found in areas such as the Southwestern United States and northern Mexico. *Lobosporangium transversal* and *Meliniomyces bicolor* are also seen to have above the average number of unique peptides in some honeybee samples (Figure 7.2). This appears to be unusual as one is rare and the other is slow-growing and associated with ectomycorrhizal root tip samples. It would be the easy path to ignore this discovery, but the number of unique peptides indicate that this is accurate detection and identification. It would appear that this discovery would be worth further consideration and study.

Another interesting fungus is *Exserohilum turcica* as seen in Figure 7.16. In this example, the unique peptides range from 0 to 25. The national average for *E. turcica* is 7.4 unique peptides. *E. turcica* is formerly known as *Helminthosporium turcicum* and is the causal agent of NCLB.

Fusarium fujikuroi is an ascomycete that is associated with diseases in corn, rice and other agricultural crops. There are many species of *Fusarium* which can share as much as 90% similarity among their sequences and are a common contaminant and well-known plant pathogens. *F. fujikuroi* has a range of 0–22 unique peptides as seen in Figure 7.20. The national average is 7.6 unique peptides.

Talaromyces atroroseus has been detected and identified as seen in Figure 7.30. This is interesting in that it was just introduced as a new species in 2013 and seen here as a fungus collected by honeybees. This fungus is found in soil and dust and is known for the ability to produce a stable red pigment for use in industry. In the selected samples, the unique peptides range from 0 to 15 and has a national average of 7.0.

Figure 7.39 is the chart for the national and regional averages for fungi. Some regions are below the national average and some are above. Overall, the national and regional averages represent a base line. It could be expected that a distressed colony would be above the national average. In this manner samples can be compared with the national average and an assessment made on the relative condition of the colony. It could be expected that a colony with fungi near the average would be in better condition than one found with a high profile. It would be a simple task to collect information from a single hive over the course of days, weeks, months and years to follow the seasonal changes. Effects of the diurnal and environmental cycles could be followed as they are expressed by the number of unique peptides per fungus. In this manner, seasonal variations could be recorded and analyzed to determine any statistically relevant events that may have either an advantageous or detrimental effect on the honeybees.

During the collection of honeybees, it was possible to identify five clear regions for further analysis. Since the same fungi data group was used for all the samples, it was possible to compare the results side by side and determine an average for each area that corresponded to the national average. In those regions where "other" fungi were noticed, this could be used to identify a difference between regions and in this manner further characterize the honeybee.

California Regional Average. The first region examined was California. The average for fungi for California tracks almost exactly with the national average for fungi, see Figure 7.34.

Florida Regional Average. The second region examined was Florida. Figure 7.35 illustrates the average for fungi on Florida honeybees when compared with the national average for fungi.

Idaho Regional Average. Figure 7.36 illustrates the average for fungi on Idaho honeybees when compared with the national average for fungi.

Iowa Regional Average. The fourth region was honeybees from Iowa. Figure 7.37 illustrates the average for fungi on Iowa honeybees when compared with the national average for fungi.

Montana Regional Average. The fifth region was honeybees from Montana. Figure 7.38 illustrates the average for fungi on Montana honeybees when compared with the national average for fungi.

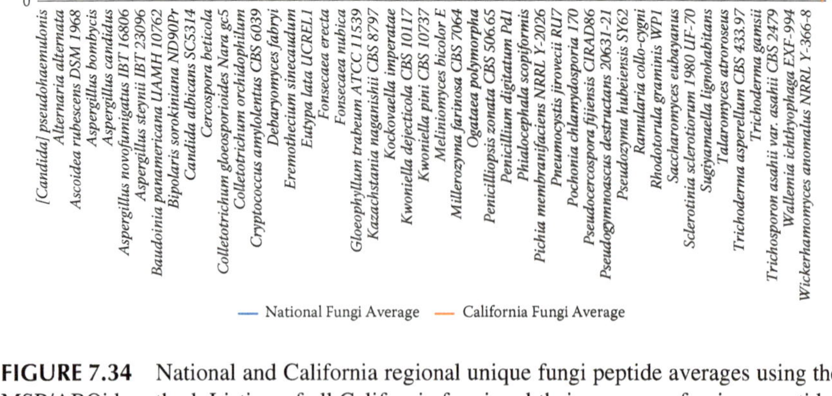

FIGURE 7.34 National and California regional unique fungi peptide averages using the MSP/ABOid method. Listing of all California fungi and their average of unique peptides is given in Appendix H.

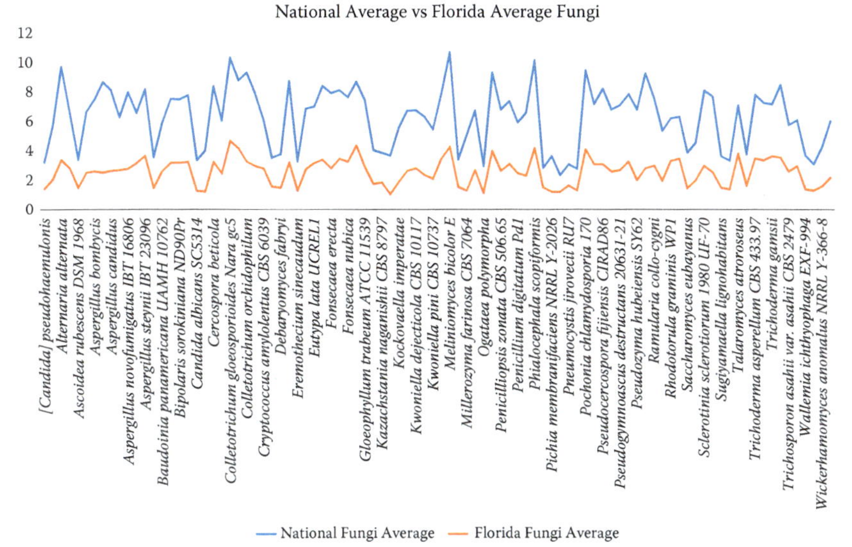

FIGURE 7.35 National and Florida regional unique fungi peptide averages using the MSP/ABOid method. Listing of all Florida fungi and their average of unique peptides is given in Appendix I.

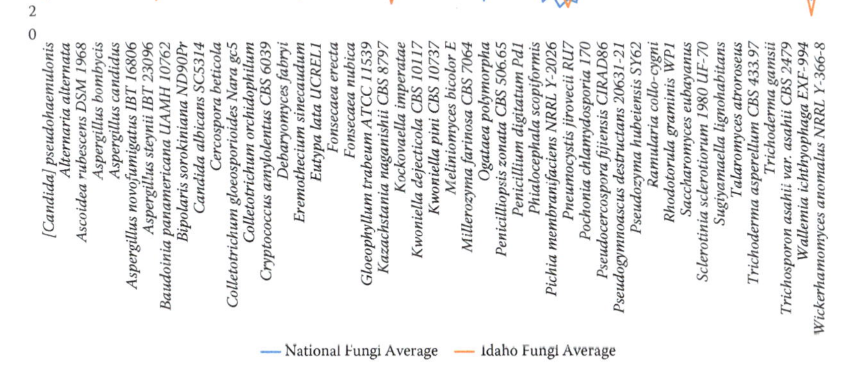

FIGURE 7.36 National and Idaho regional unique fungi peptide averages using the MSP/ABOid method. Listing of all Idaho fungi and their average of unique peptides is given in Appendix J.

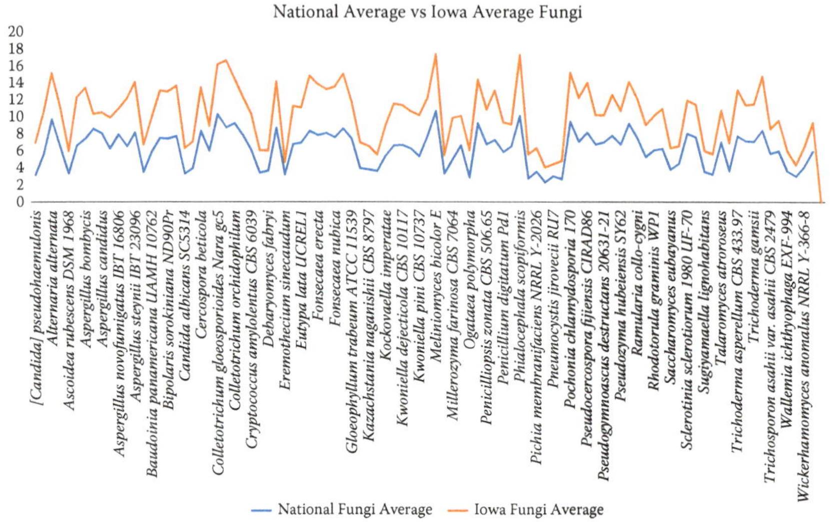

FIGURE 7.37 National and Iowa region unique fungi peptide averages using the MSP/ABOid method. Listing of all Iowa fungi and their average of unique peptides is given in Appendix K.

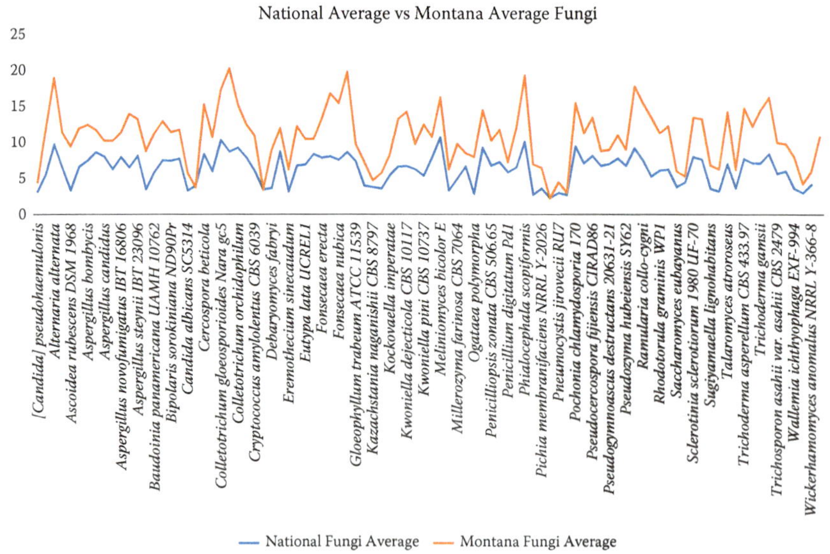

FIGURE 7.38 National and Montana region unique fungi peptide averages using the MSP/ABOid method. Listing of all Montana fungi and their average of unique peptides is given in Appendix L.

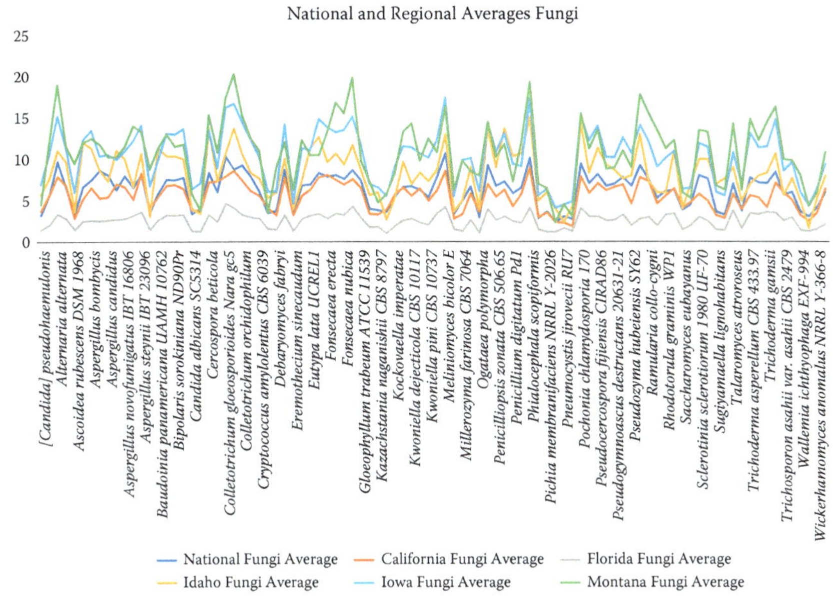

FIGURE 7.39 National and Regional averages for unique fungi peptides.

7.3 DISCUSSION

Several hundred fungi have been identified. Many are bacteria with symbiotic relationships with plants and others are soil inhabitants. The national Average for fungi ranges from less than 2 to 10 unique peptides for these fungi. Overall, 31 species are discussed in detail and of these 8 species have the highest averaging from 8 to 11 unique peptides. These are further classified, as several being symbiotic mycorrhizal fungi, others being bioremediation fungi and the others being soil inhabitants. Ascomycetes top the list in number of microbes identified.

The classification of fungi is ongoing, and some of them may be under-speciated at the moment, because there are many not sequenced and remain unnamed. The rapid increase in sequencing fungi will contribute to identification and cement phylogenic relationships. Nevertheless, in this chapter, hundreds of fungi have been detected and discussed that were collected by honeybees. Some of these fungi may have been collected from forging others or may have been collected from the hive itself.

Environmentalists are encouraged to study these fungi and compare them with the bacteria and viruses to determine the mutual relationships and assess these relationships within a region, area or a locality.

The number of fungi species that can be identified is continuing to increase due to the increasing number of sequences added to the NCBI. More than 250,000 prokaryotic microbes have been sequenced, but only 6,682 are fungi. This number increases daily.

The ability to add new microbes has been discussed in Chapter 2, but it remains important to remember that as new fungi are sequenced and added to the fungi data group, the samples that have been analyzed can be re-analyzed using the old computer files. In this manner there is no need to collect a new sample, and large archived sample sets can be re-examined and a search can be made for the new fungi.

Finally, considering all the fungi species that exist and are being discovered, it is possible to consider that we live in a very rich environment of microbes of which fungi are a large part. There is so much potential for this type of information collected by the honeybees. In their regular work, they have the information about their working environment with them, all we have to do is sample, detect and identify the microbes. Microbes of all sorts are found with the honeybees, and the diversity of these microbes may be useful in measuring or monitoring changes to the environment.

8 *Nosema*

One microbe stands out from the others and is a keen interest to beekeepers, and that is the microsporidian *Nosema ceranae*. It is a small, unicellular parasite that is known to affect *Apis sp.* It is of the kingdom fungi and of the phylum Microsporidia. It has a durable spore-forming stage allowing it to remain viable for several years. First reported in 1996 and reported as a disease in 2004, it has been associated with honeybees (*Apis mellifera*) in the United States and elsewhere and has become a microbe of interest. As a result, the genome sequence of *Nosema ceranae* was downloaded from the National Center for Biotechnology Information (NCBI) and added to a data group. The resulting detection and identification of *Nosema ceranae* was made, and a national average and five regional averages (California, Florida, Idaho, Iowa and Montana) were determined.

8.1 NOSEMA

Considering the recent identification of this fungus and the recent apparent spread to nearly every beekeeping operation and the high national and regional averages, it can be seen that this is a microbe of interest. Much discussion has been made about the relationship of *Nosema ceranae* and other microorganisms in relation to honeybee health (Bromenshenk et al., 2010).

When hundreds of microbes identified and exemplified in the various chapters are compared with *Nosema*, it may be possible to identify new associations between bacteria, viruses and other fungi to further our understanding of this particular fungus and honeybees.

8.2 NATIONAL AND REGIONAL AVERAGES FOR *NOSEMA CERANAE*

The national Average for *Nosema ceranae* is 300 unique peptides. The California average is 261 and Florida 177. Idaho has 232 unique peptides as an average, followed by 590 for Iowa and the largest average of 792 for the Montana region (Figure 8.1).

8.3 DISCUSSION

Nosema ceranae is pathogenic to *Apis mellifera* and in association with other microorganisms may be a continuing and serious threat to honeybees. The apparent rapid spread of this microsporidian rises concerns about other possible pathogens that may emerge and threaten honeybees.

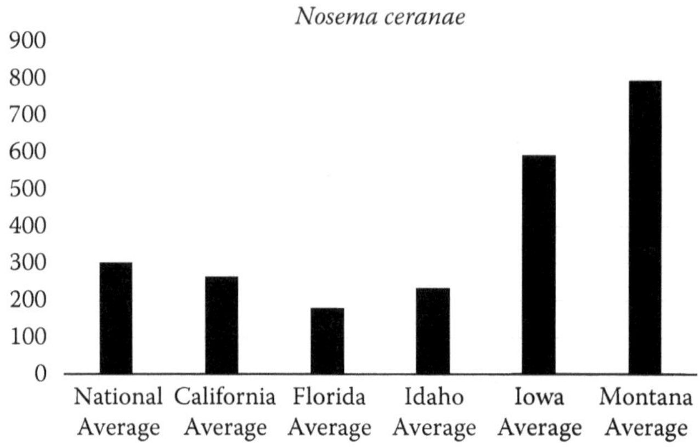

FIGURE 8.1 National and regional averages for *Nosema ceranae.*

Nosema can be counted and monitored by optical microscopy. The ability to detect and identify the fungus by mass spectrometry proteomics (MSP)/ABOid methods furthers this capability by adding *Nosema* to the reported other microbes, all determined from a single sample.

The classification of fungi is ongoing, and some of them may be underspeciated at the moment because there are many not sequenced and remain unnamed. Additional species of *Nosema* or *Nosema*-like fungi can be expected. The rapid increase in fungi sequencing will contribute to identification and cement phylogenic relationships. Nevertheless, hundreds of fungi have been detected and identified that were collected by honeybees, among them *Nosema ceranae.* Close attention needs to be directed as to the source of *Nosema* infections, treatment effectiveness and distribution of the fungi, which will be useful in the control of this pathogen. Some of these fungi may have been collected from foraging while others may have been collected from the hive itself.

Environmentalists are encouraged to study the *Nosema* results and compare them with the bacteria, viruses and other fungi to determine mutual relationships and assess these relationships within a region, area or a locality.

The ability to add new microbes has been discussed in Chapter 2, but it remains important to remember that as new fungi are sequenced and added to the fungi data group or to the *Nosema* data group, archived samples can be reanalyzed using the old computer files. In this manner there is no need to collect a new sample, and large archived sample sets can be re-examined and a search can be made for the new fungi.

The number of fungi species that can be identified continues to increase due to the increasing number of sequences added to the NCBI. More than 250,000 prokaryotic microbes have been sequenced of which 6,682 are fungi. This number increases daily.

Finally, considering all the species of microbes that exist and are continually being discovered, it is possible to consider beginning a detailed analysis of the microbiome in which we live. There is so much potential for this type of information collected by the honeybees. In their regular work, honeybees have collected the information about their working environment, all we have to do is sample, detect and identify the microbes. Microbes of all sorts are found with the honeybees, and the diversity of these microbes is useful in measuring or monitoring changes to the environment. This is of particular interest considering the abundance of *Nosema ceranae* in the samples. The continued spread of *Nosema ceranae* may lead to effective control and effects of different environments on the fungus.

9 Viruses

Several hundred viruses collected by honeybees were identified. Of these viruses, 103 were detected frequently enough to make an average. Of these 103, 14 were identified as standing above all the rest as seen in Figure 9.1, the national average for viruses. Five viruses can be grouped as belonging to the Herpesviridae family, eight belong to the Poxviridae family, and the African swine fever virus stands alone. For all the viruses detected at low levels, we have three families of viruses that dominate: herpes, pox and African swine fever. It is beyond the scope of this book to analyze "why" these particular viruses occur, but the fact that they do occur is equally important. Figure 9.1 illustrates the frequency of these groups as shown with the national average and the averages for California, Florida, Idaho, Iowa and Montana for just these major groups. Iowa and Montana averages are above the national average. California, Florida and Idaho are all about the same as the national average. Camelpox, monkeypox and vaccinia viruses show greater numbers of peptides than Cercopithecine herpesvirus 5. As to the meaning of these findings, well it is clear that these viruses are being detected and identified. Finding more than 80 unique peptides for camelpox virus is interesting, as it is for monkeypox and vaccinia. It would appear that Iowa and Montana have a greater repository of these viruses than the other regions. The other three regions as well as the national average would indicate a range of approximately 10–20 unique peptides.

The major three groups and the high number of unique peptides found in these groups should not distract from the detection and identification of the other viruses. Any of these other viruses could become a detection of interest with a change in the source for the honeybee collectors. Newcastle disease virus has an average in the Montana region of over seven unique peptides as well as Western equine encephalitis virus and others.

The following sections will discuss the major virus groups and expand on the relationships between the national average of viruses collected by honeybees and the various regions. It may be important and useful to monitor the viruses collected by honeybees as an indication of a potential source of new emerging concerns.

9.1 GENERAL VIRUS GROUP

The first thought when looking over the detection and the identification is the wide variety of viruses detected. Figure 9.1 illustrates this diversity and gives a national average for the 103 viruses frequently detected.

Several other thoughts come to mind when looking over these viruses. The main one is why? Why are we seeing these in our samples and why are they so

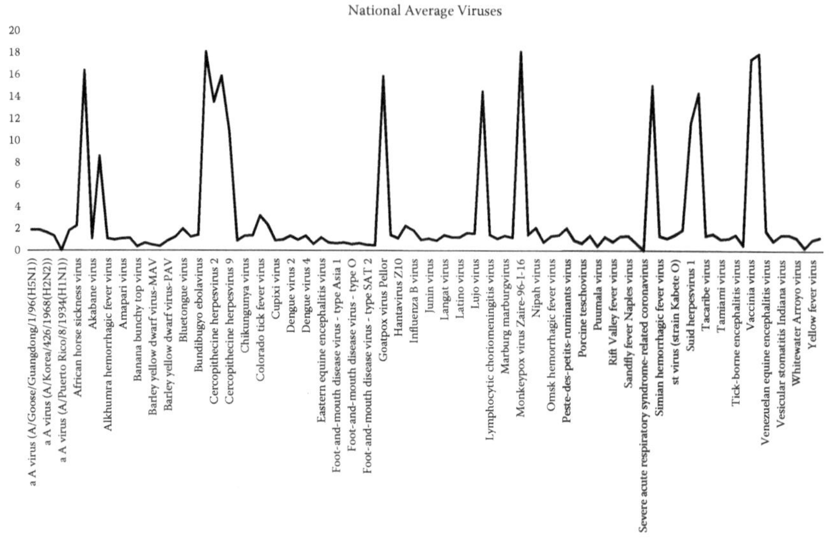

FIGURE 9.1 National average for viruses. Listing of all national viruses and their average of unique peptides is given in Appendix M.

widely distributed? As you read the following text, you will see that these groups have in common a wide distribution and a history of a hardy survivability. This would indicate that the pox viruses, the herpesviruses and evidently the African swine fever virus can survive in a wide distribution. All the regions show these viruses, there is no doubt they are here.

The final thoughts on this issue are two: "what do I do about them" and "what does this mean to me"? First, these viruses have been around a long time. They are detected at low levels. We have not sustained, as far as we know, any major negative side effects. We consume them, we breathe them and we live with them. The population has immunity from many microbes, and as a result they are not a present threat to the population. As further evidence, it appears that most microbes are not a direct threat to us because we have not sustained an outbreak of any of these viruses among the population. This means that although we have detected these viruses, not all of them are a threat to us and we should treat their detection not as a worry but to that of interesting information. The natural microbe population should be studied as there may be pathogens lurking. Tetanus, polio, colds, and other microbes are in this microflora, and weakened or immunocompromised individuals may be at risk. While we could see the benign natural occurring microflora that surrounds us all, it also is evident that we are detecting pathogenic microbes as well. Monitoring these pathogenic microbes that have the potential of becoming a threat is useful considering that some of these microbes may be just waiting to pounce if they have the chance. Outbreaks of infection that lead to a pandemic are a different matter. In these cases, such as influenza and COVID-19, it would be helpful to know where the infectious virus (microbe) is located in the environment so they can be controlled and eliminated.

The national average for viruses reveals several viruses that are more or less common in our samples. There are 20 viruses that stand out as having a greater average of unique peptides. These are listed below with a brief description as to their averages and descriptions.

9.1.1 African Swine Fever Virus (ASFV) – *Variola Porcina*

African swine fever virus (ASFV) is the only species in the order Asfuvirales, family Asfarviridae and genus *Asfiviru. Variola porcina* is a large DNA virus of the phylum Nucleocytoviricota. It is a large, icosahedral, double-stranded DNA virus with a linear genome of 189 kilobases containing more than 180 genes (Linda K. Dixon, 2008).

Although ASFV does not cause disease in humans, it has been a problem in swine. It was thought to have evolved around 1700. It can be spread by ticks and pigs and also by food products that contain the virus. It causes hemorrhagic fever in pigs usually within a week of infection.

Given this information, a question then remains, why are we seeing it among the prominent viruses isolated from honeybees? This virus is present everywhere across the United States as evidenced by the analysis of honeybee collections. One thought on this is that following the widespread pig farming in the early 1900s and the first major outbreaks of ASFV at that time to sporadic outbreaks from time to time the virus became endemic throughout much of Africa, Europe and when it crossed the Atlantic in the Caribbean. In 2018 the virus spread to Asia. Since the virus can remain in an infectious state in the *Ornithodoros* tick vector for months or up to years, it is likely that the virus has spread among the natural hosts, warthogs, bushpigs and soft ticks. The continued spread of this virus since being first noticed to recent times would indicate a rational for having ASFV picked up and then isolated from honeybees.

Also associated with pigs is Aujeszky's disease (also known as pseudorabies). This is a viral disease of pigs and endemic in most parts of the world. It is caused by *Suid* herpesvirus , a member of the subfamily Alphaherpesvirinae and the family Herpesviridae. The virus infects a variety of mammals, but only pigs are able to survive a productive infection and are thereby considered the natural host (Kluge, Beran, & Platt, 1999).

The national average for ASFV is 16.4 unique peptides, Figure 9.2. The averages for California, Florida, Idaho, Iowa and Montana are shown in Figure 9.12.

9.1.2 Alcelaphine Herpesvirus 1 (AlHV-1)

Alcelaphine herpesvirus 1 (AlHV-1) is a large DNA virus of the phylum Peploviricota, also known as malignant catarrhal fever. *Alcelaphine herpesvirus*

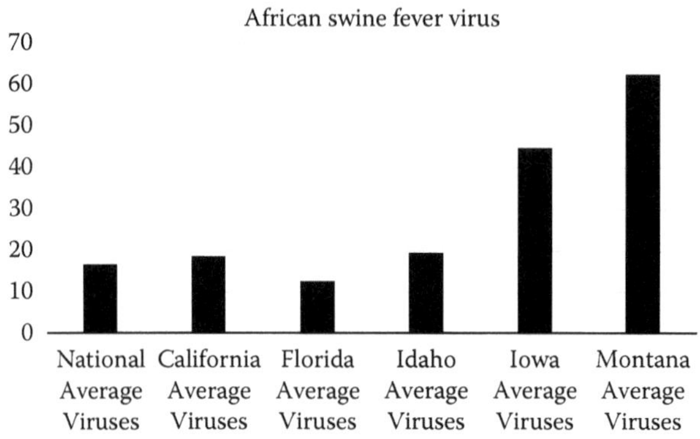

FIGURE 9.2 Detection and identification of African swine fever irus using the MSP/ABOid method.

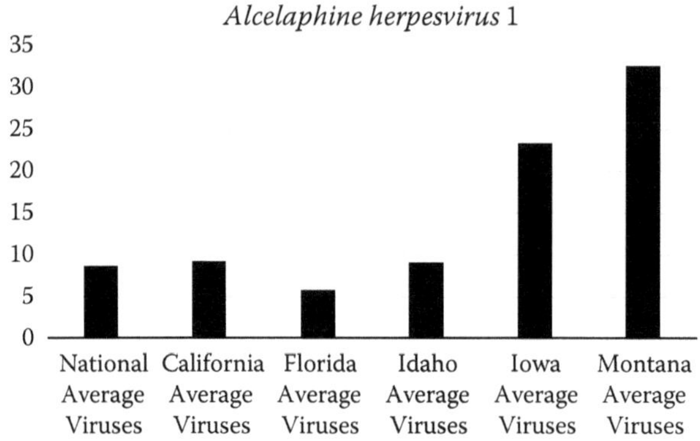

FIGURE 9.3 Detection and identification of *Alcelaphine herpesvirus* using the MSP/ABOid method.

1 serves as the prototype virus of the *Macavirus* genus of the Gammaherpesvirinae family Herpesviridae. Bovine malignant catarrhal fever (BMCF) is a fatal lymphoproliferative disease (Toole & Li, 2014).

The national average for AlHV-1 is 8.6 unique peptides, (Figure 9.3). The averages for California, Florida, Idaho, Iowa and Montana are shown in Figure 9.12.

9.1.3 Camelpox Virus (CMLV)

Camelpox virus is a large enveloped DNA virus that is taxonomically assigned to the family Poxviridae, subfamily Chordopoxvirinae, and genus *Orthopoxvirus*. Other members of the genus (i.e., the orthopoxviruses) include important human pathogens such as variola (smallpox), monkeypox, cowpox and vaccinia viruses, in

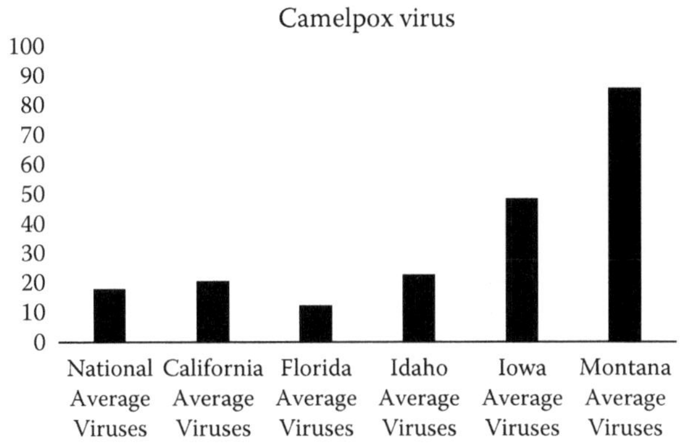

FIGURE 9.4 Detection and identification of camelpox virus using the MSP/ABOid method.

addition to those of lesser importance such as ectromelia, raccoonpox, skunkpox, taterapox and volepox (Moss, 2013). Camelpox (CMLV) is a contagious viral disease of camels that occurs throughout the camel-breeding countries of northern Africa, the Middle East and Asia (Balamurugan et al., 2013). Like other poxviruses, camelpox virions show a high degree of environmental stability and can remain infectious over several months (Rheinbaden, Gebel, Exner, & Schmidt, 2007).

The national average for CMLV is 18.1 unique peptides (Figure 9.4). The averages for California, Florida, Idaho, Iowa and Montana are shown in Figure 9.12.

9.1.4 Cercopithecine Herpesvirus 5 (CeHV-5)

African green monkey cytomegalovirus (CMV) and *Cercopithecine herpesvirus* 5 (CeHV-5) are viruses in the genus *Cytomegalovirus*, subfamily Betaherpesvirinae, family Herpesviridae, and order Herpesvirales. African green monkeys (*Chlorocebus* spp.) serve as natural hosts.

The national average for CMLV is 15.9 unique peptides (Figure 9.5).

The averages for California, Florida, Idaho, Iowa and Montana are shown in Figure 9.12.

9.1.5 Goatpox Virus Pellor (GTPV)

Goatpox virus Pellor, also known as GTPV strain Pellor (PL), is a contagious viral disease which affects goat and sheep populations. The virus is endemic in southwestern Asia, India and northern and central Africa. GTPV is one of three recognized members of the genus *Capripoxvirus*.

The national average for is GTPV 15.8 unique peptides, Figure 9.6. The averages for California, Florida, Idaho, Iowa and Montana are shown in Figure 9.12.

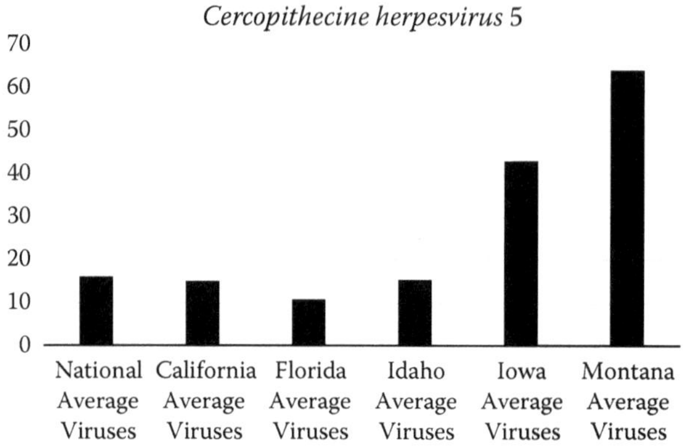

FIGURE 9.5 Detection and identification of *Cercopithecine herpesvirus* 5 using the MSP/ABOid method.

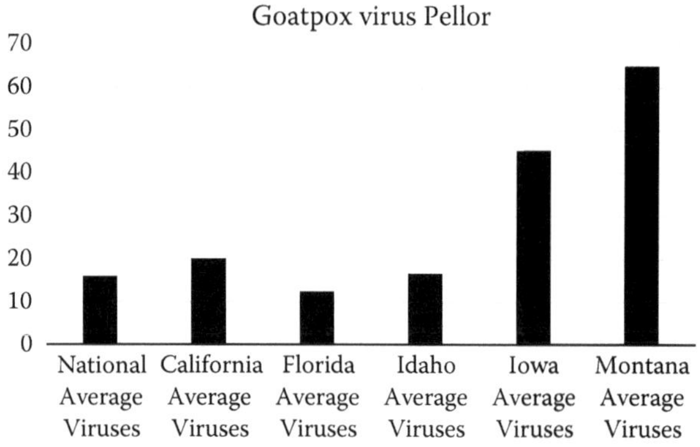

FIGURE 9.6 Detection and identification of goatpox virus Pellor using the MSP/ABOid method.

9.1.6 Lumpy Skin Disease Virus (LSD)

Lumpy skin disease virus (LSDV) is double-stranded DNA virus and a member of the *capripoxvirus* genus of *Poxviridae.* Capripoxviruses (CaPVs) represent one of eight genera within the Chordopoxvirus (ChPV) subfamily including sheeppox virus and goatpox virus. The capripoxviruses are brick-shaped and are different than orthopoxvirus virions in that they have a more oval profile and are average in size of 320 nm × 260 nm (Tulman et al., 2001).

Lumpy skin disease (LSD) is an infectious disease in cattle. It has spread rapidly through the Middle East, southeast Europe, the Balkans, Caucasus, Russia and Kazakhstan (World Animal Health Information Database (WAHID) Interface, 2020) and is considered to be one of the emerging threats to Europe and Asia (Allepuz, Casal, & Beltrán-Alcrudo, 2019; Machado et al., 2019).

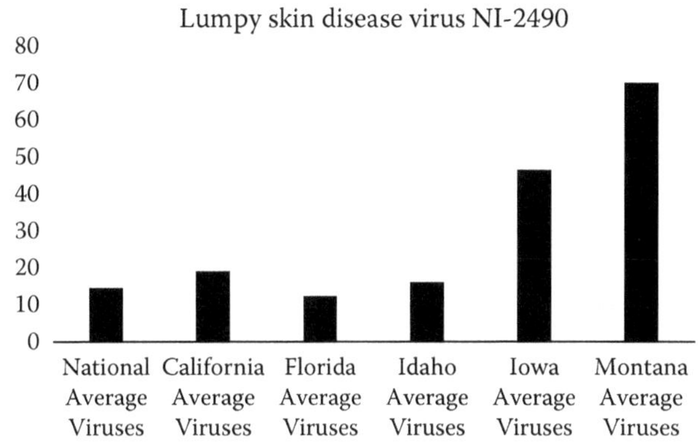

FIGURE 9.7 Detection and identification of lumpy skin disease virus using the MSP/ABOid method.

LSDV mainly affects cattle and is also been seen in giraffes, water buffalo and impalas and other animals. Fine-skinned Bos taurus cattle breeds such as Holstein-Friesian and Jersey are the most susceptible to the disease.

The national average for LSD is 14.5 unique peptides (Figure 9.7). The averages for California, Florida, Idaho, Iowa and Montana are shown in Figure 9.12.

9.1.7 Monkeypox Virus Zaire-96-I-16 (MPV)

Monkeypox virus (MPV) is a double-stranded DNA, zoonotic virus and a species of the genus *Orthopoxvirus* in the family Poxviridae. It is one of the human orthopoxviruses that includes variola (VARV), cowpox (CPX), and vaccinia (VACV) viruses. But MPV is neither a direct ancestor to, nor a direct descendant of, the variola virus which causes smallpox (Breman et al., 1979).

It is interesting that monkey pox is associated with monkeys but they are not the main reservoir of the virus. Antibodies have been found in a variety of animals (Khodakevich, Ježek, & Kinzanzka, 1986).

The MPV has been found in ground squirrels and they may be a reservoir of the virus (Sergeev et al., 2017). This would lead to the suggestion that MPV may be much more prevalent in the environment then commonly thought and that the honeybees are picking it up just as they would any other virus in their activity.

The national average for MPV is 18.1 unique peptides (Figure 9.8).

The averages for California, Florida, Idaho, Iowa and Montana are shown in Figure 9.12.

9.1.8 Sheeppox Virus

The national average for SPV is 15 unique peptides (Figure 9.9).

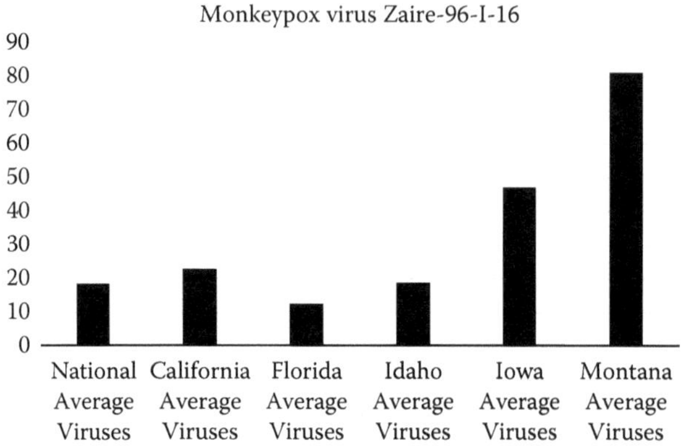

FIGURE 9.8 Detection and identification of monkeypox virus using the MSP/ABOid method.

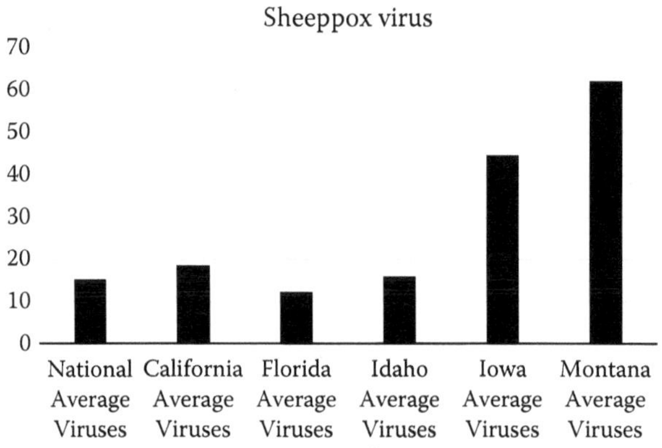

FIGURE 9.9 Detection and identification of sheeppox virus using the MSP/ABOid method.

The averages for California, Florida, Idaho, Iowa and Montana are shown in Figure 9.12.

See goatpox in Figure 9.6.

9.1.9 Vaccinia Virus (VACV)

Vaccinia viruses are a close relatives of the smallpox virus (VARV) and are also pathogenic to humans. These include the old world orthopoxviruses, vaccinia (VACV), cowpox (CPXV) and monkeypox (MPXV). Rodents are the major natural reservoir of cowpox and monkeypox (Moss, 2007; Shchelkunov, 2005).

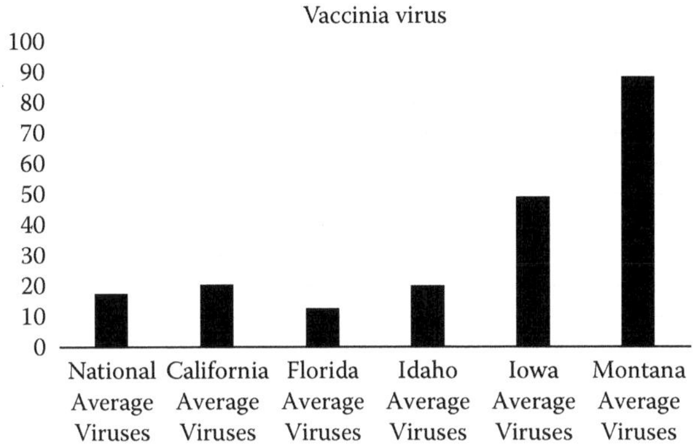

FIGURE 9.10 Detection and identification of vaccinia virus using the MSP/ABOid method.

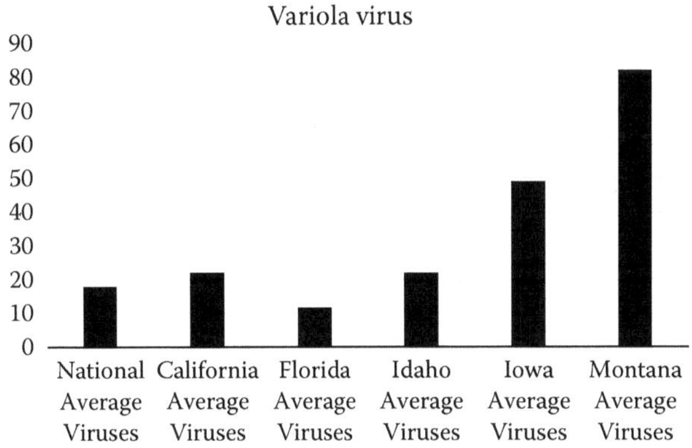

FIGURE 9.11 Detection and identification of variola virus using the MSP/ABOid method.

The national average for vaccinia is 14.5 unique peptides (Figure 9.10). The averages for California, Florida, Idaho, Iowa and Montana are shown in Figure 9.12.

9.1.10 Variola Virus (VARV)

Variola is a large brick-shaped virus measuring approximately 302–350 nm × 244–270 nm, with a single linear double-stranded DNA genome 186 kilobase pairs (kbp) in size and containing a hairpin loop at each end (Roossinck, 2016).

Smallpox is the disease caused by the variola virus (VARV), which belongs to the genus Orthopoxvirus. Although declared eradicated from the human community by the World Health Organization (WHO) in 1980, the WHO, did not, however,

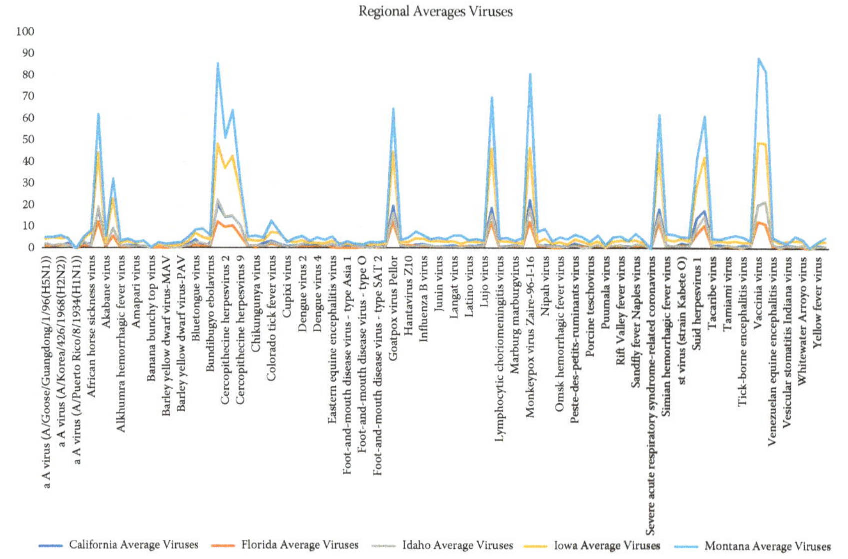

FIGURE 9.12 Regional averages for viruses. Listing of all viruses and their average of unique peptides for each region is given in Appendices N–R.

remove it from the list of potential and dangerous pathogens, since the virus remains in laboratories for study and it has been known to reoccur throughout history. This possibility is furthered by the fact that people have not regularly been vaccinated against VARV for a long time and a susceptible and vulnerable population has grown making a re-emergence of the virus possible (Singh, Balamurugan, Bhanuprakash, Venkatesan, & Hosamani, 2012).

Nevertheless, it is not the purpose of this book to explore these options, but rather only reports on the collection of VARV from the environment by honeybees. Detected at low levels, but still detected and at such a level to separate itself from the more than 500 other viruses found in the background. One thought on this matter is that the honeybees can be useful in monitoring the level of VARV in a given environment.

The national average for VARV is 14.5 unique peptides (Figure 9.11). The averages for California, Florida, Idaho, Iowa and Montana are shown in Figure 9.12.

9.2 DISCUSSION

More than a hundred viruses have been identified. Many viruses are common in the environment and some have relationships with plants and others are soil inhabitants, some are pathogenic. The national average for viruses ranges from less than 2 to 18 unique peptides. In this chapter, nine viruses are discussed in detail because they stand out from the field by having 12–18 unique peptides (Figure 9.13).

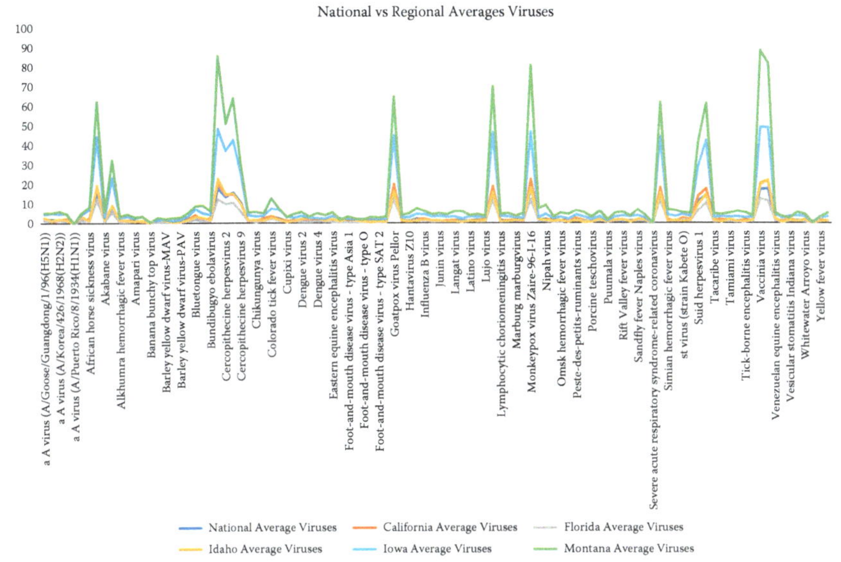

FIGURE 9.13 National and regional averages for viruses.

The classification of viruses is ongoing, and some of them may be under-speciated at the moment because many are not sequenced and remain un-named. Viruses have a high mutation rate, and the number of species or strains within a genus and species can rapidly change. The rapid increase in sequencing viruses will contribute to identification and cement phylogenic relationships. Nevertheless, in this chapter over a hundred viruses have been detected and identified that were collected by honeybees with only nine standing out from the field. Some of these viruses may have been collected from foraging others may have been collected elsewhere or from the hive itself.

Environmentalists are encouraged to study these viruses and compare them with the bacteria and fungi to determine the mutual relationships and assess these relationships within a region, area or a locality.

The number of virus species that can be identified is continuing to increase due to the increasing number of sequences added to the NCBI. More than 37,000 viruses have been sequenced; the number increases daily.

The ability to add new microbes has been discussed in Chapter 2, but it remains important to remember that as new viruses are sequenced and added to the virus data group, the samples that have been analyzed can be re-analyzed using the old computer files. In this manner there is no need to collect a new sample, and large archived sample sets can be re-examined and a search can be made for the new viruses. An example of this is demonstrated in Chapter 5, where the SARS and COVID-19 viruses were downloaded into their own data group. Recently over 450 new viruses were added to the virus data group.

Finally, considering all the virus species that exist, are being discovered and exist everywhere, it is possible to consider that we live in a very rich environment of microbes of which viruses are a large part. There is so much potential for the type of information collected by the honeybees. In their regular work, honeybees collect all the microbe information about their working environment. They carry the microbes with them; all we have to do is sample, detect and identify. Microbes of all sorts are found with the honeybees, and the diversity of these microbes may be useful in measuring or monitoring changes to the environment.

10 Water Microbes (EPA Standards)

10.1 WATER MICROBES

10.2 NATIONAL AVERAGE OF WATER MICROBES

The national average figure (Figure 10.1) has 10 microbes with higher unique peptide averages than others and forms a distinctive pattern for their distribution. These 10 microbes are described further.

10.2.1 *CAMPYLOBACTER JEJUNI*

Campylobacter jejuni is in a genus of bacteria that is among the most common causes of bacterial infections in humans worldwide and is one of the most common causes of food poisoning in Europe and in the United States (Foodsafety.gov, 2020).

Campylobacter means "curved rod", deriving from the Greek *kampylos* (curved) and *baktron* (rod). *C. jejuni* is also commonly found in animal feces. *Campylobacter* is a helical-shaped, non-spore-forming, gram-negative, microaerophilic, non-fermenting, motile bacterium with a single flagellum at one or both poles (Balaban & Hendrixson, 2011), which are also oxidase-positive and grow optimally at 37–42°C. (Ruam, 2004). Of its many species, *C. jejuni* is considered one of the most important from both a microbiological and public health perspective (Blaser, 1997).

Campylobacter organisms have a large animal reservoir, with up to 100% of poultry, including chickens, turkeys and waterfowl, having asymptomatic intestinal infections. The major reservoirs of *C. fetus* are cattle and sheep.

The national average for *C. jejuni* is 14.5 unique peptides (Figure 10.1). The averages for California, Florida, Idaho, Iowa and Montana are shown in Figure 10.7.

10.2.2 *ESCHERICHIA COLI* O157

There are many strains of *Escherichia coli* which are gram-negative and oxidase-negative, facultative anaerobic, rod-shaped bacteria that are part of the normal intestinal flora and grow easily in most culture media. *E. coli* is classified into between 150 and 200 serotypes, and is the most common cause of *E. coli* diarrhea in farm animals.

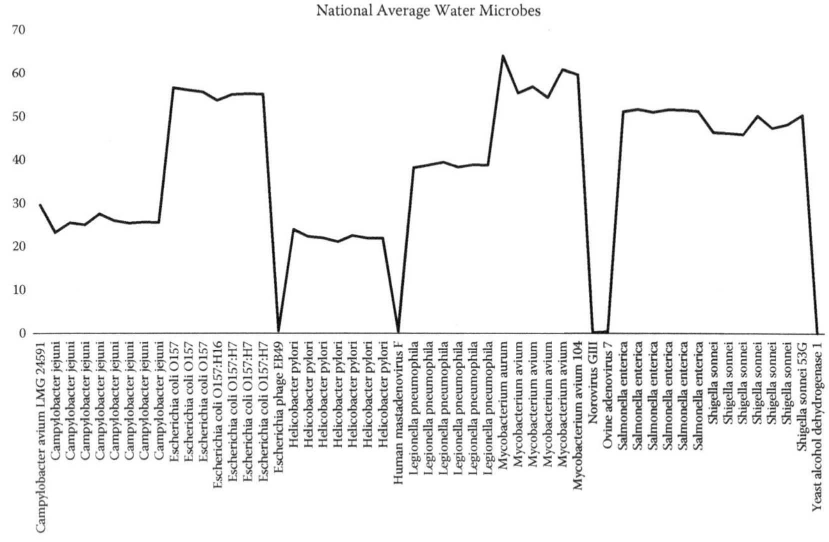

FIGURE 10.1 National average for water microbes.

While it is relatively uncommon, the *E. coli* serotype O157:H7 can naturally be found in the intestinal contents of some cattle, goats and even sheep (Fairbrother, 1999). The digestive tract of cattle can be asymptomatic carriers of the bacterium (Pruimboom-Bre IM, 2000). The prevalence of *E. coli* O157:H7 in North American feedlot cattle herds ranges from 0 to 60% (Jeon, Elzo, DiLorenzo, Lamb, & Jeong, 2013). Some cattle may also be so-called super-shedders of the bacterium. Super-shedders pass the bacterium in feces. Although super-shedders constitute a small proportion of the cattle in a feedlot, they may account for >90% of all *E. coli* O157:H7 excreted (Chase-Topping, Gally, Low, Matthews, & Woolhouse, 2008).

Considering the large number of strains of *E. coli*, it is not surprising that the honeybees would collect multiple strains of this bacteria.

The national average for *E. coli* is 14.5 unique peptides (Figure 10.1). The averages for California, Florida, Idaho, Iowa and Montana are shown in Figure 10.7.

10.2.3 *Helicobacter pylori*

Helicobacter pylori, previously known as *Campylobacter pylori*, is a gram-negative, helical-shaped, microaerophilic bacterium usually found in the stomach (Yamaoka, 2008).

H. pylori infection is very prevalent, and is likely present in the gastric tissues of 74% of middle-aged adults in developing countries and 58% in developed countries (Ferlay et al., 2018).

The national average for *H. pylori* is 14.5 unique peptides (Figure 10.1). The averages for California, Florida, Idaho, Iowa and Montana are shown in Figure 10.7.

10.2.4 Human Mastadenovirus F

The national average for human mastadenovirus F is 14.5 unique peptides (Figure 10.1). The averages for California, Florida, Idaho, Iowa and Montana are shown in Figure 10.7.

10.2.5 *Legionella pneumophila*

The national average for *Legionella pneumophila* is 14.5 unique peptides (Figure 10.1) The averages for California, Florida, Idaho, Iowa and Montana are shown in Figure 10.7.

10.2.6 *Mycobacterium avium*

The national average for *Mycobacterium avium* is 14.5 unique peptides (Figure 10.1). The averages for California, Florida, Idaho, Iowa and Montana are shown in Figure 10.7.

10.2.7 *Norovirus* GIII

The national average for *Norovirus* is 14.5 unique peptides (Figure 10.1). The averages for California, Florida, Idaho, Iowa and Montana are shown in Figure 10.7.

10.2.8 *Ovine adenovirus* 7

The national average for *Ovine adenovirus* is 14.5 unique peptides (Figure 10.1). The averages for California, Florida, Idaho, Iowa and Montana are shown in Figure 10.7.

10.2.9 *Salmonella enterica*

The national average for *Salmonella enterica* is 14.5 unique peptides (Figure 10.1). The averages for California, Florida, Idaho, Iowa and Montana are shown in Figure 10.7.

10.2.10 *Shigella sonnei*

The national average for *Shigella sonnei* is 14.5 unique peptides (Figure 10.1). The averages for California, Florida, Idaho, Iowa and Montana are shown in Figure 10.7.

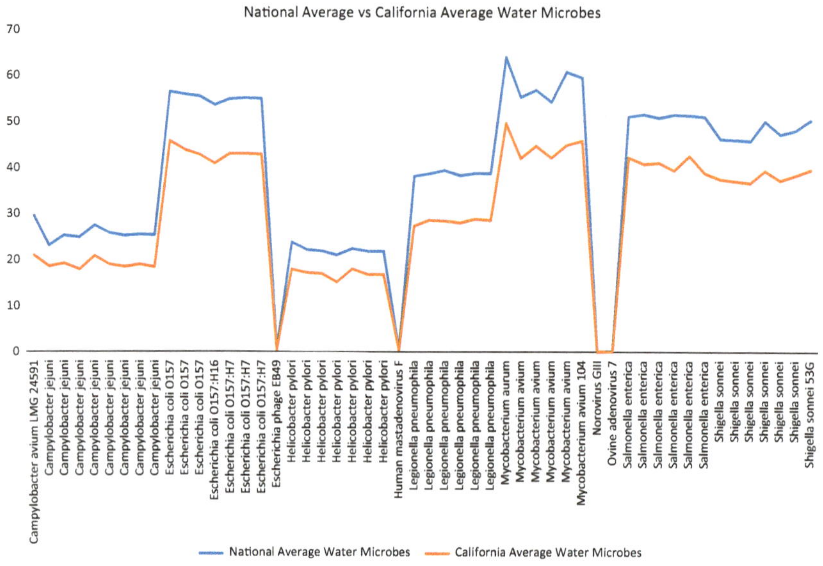

FIGURE 10.2 National and California region unique water microbe peptide averages using the MSP/ABOid method.

10.3 CALIFORNIA REGIONAL AVERAGE

Figure 10.2 contains the distribution of the water microbes for the California region which display several species of interest as discussed in section 10.1.

10.4 FLORIDA REGIONAL AVERAGE

Figure 10.3 contains the distribution of the water microbes for the Florida region which display several species of interest as discussed in section 10.1.

10.5 IDAHO REGIONAL AVERAGE

Figure 10.4 contains the distribution of the water microbes for the Idaho region which display several species of interest as discussed in section 10.1.

10.6 IOWA REGIONAL AVERAGE

Figure 10.5 contains the distribution of the water microbes for the Idaho region which display several species of interest as discussed in section 10.1.

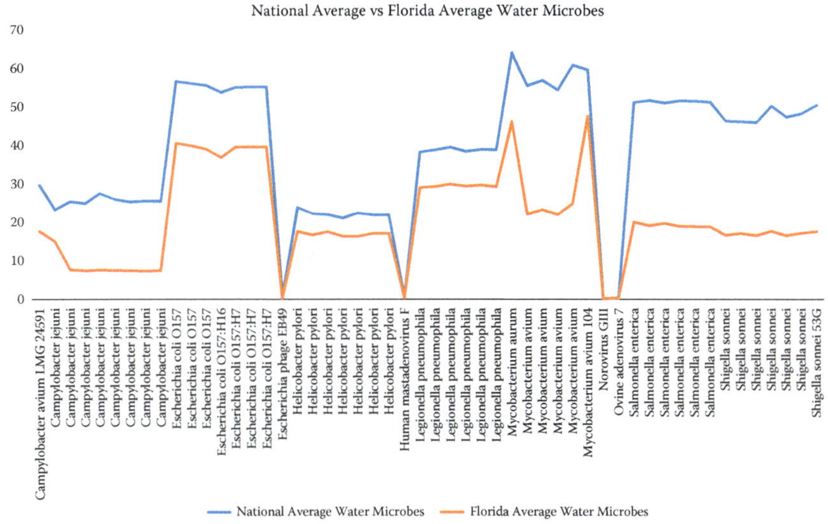

FIGURE 10.3 National and Florida region unique water microbe peptide averages using the MSP/ABOid method.

10.7 MONTANA REGIONAL AVERAGE

Figure 10.6 contains the distribution of the water microbes for the Idaho region which display several species of interest as discussed in section 10.1.

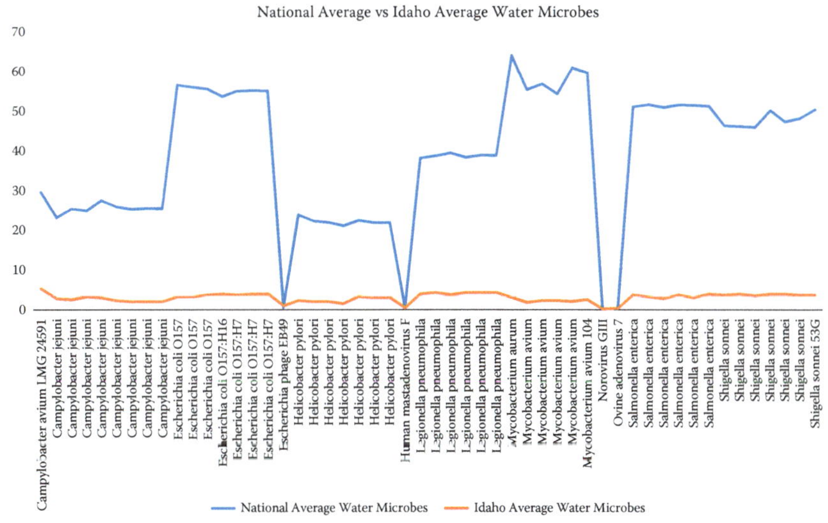

FIGURE 10.4 National and Idaho region unique water microbe peptide averages using the MSP/ABOid method.

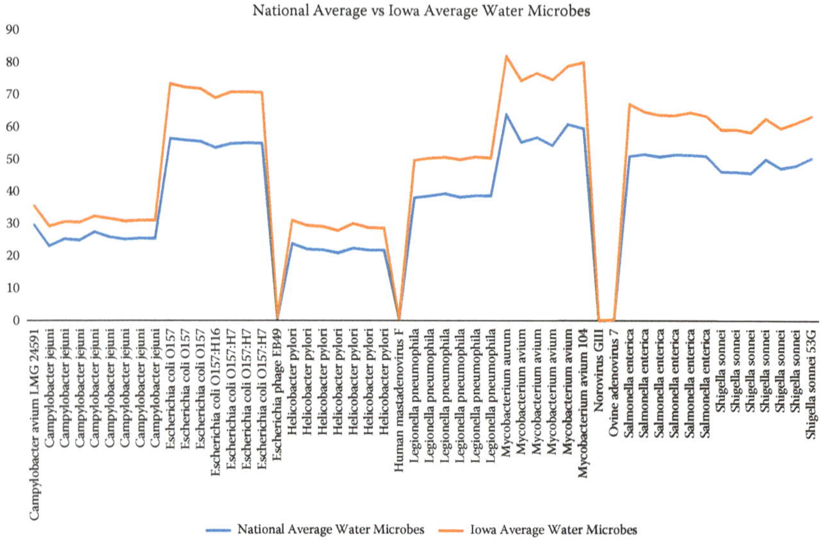

FIGURE 10.5 National and Iowa region unique water microbe peptide averages using the MSP/ABOid method.

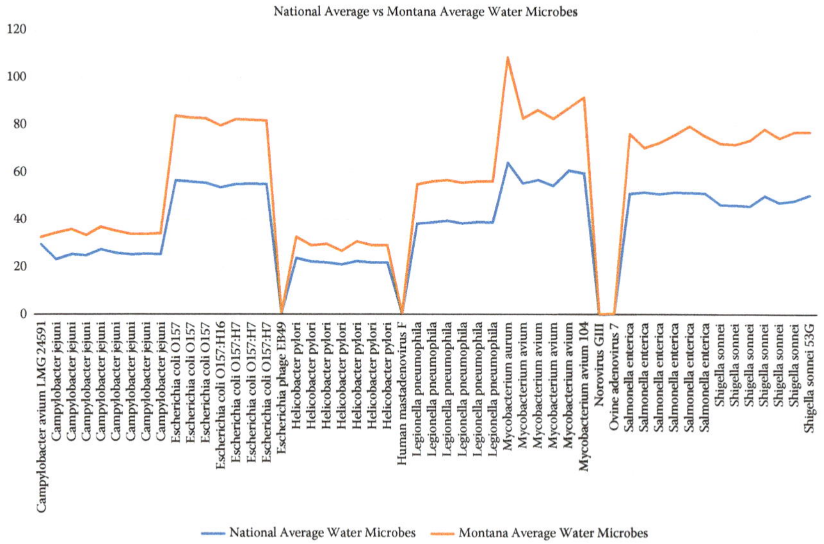

FIGURE 10.6 National and Montana region unique water microbe peptide averages using the MSP/ABOid method.

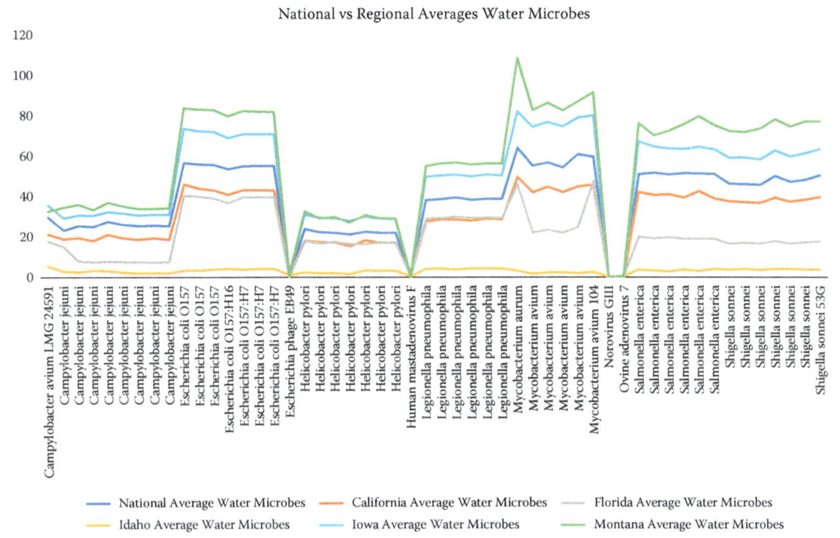

FIGURE 10.7 National and regional averages for water microbes.

10.8 NATIONAL AND REGIONAL AVERAGES

Figure 10.7 contains the distribution of the water microbes for all the regions versus the national average which display several species of interest as discussed in section 10.1.

10.9 DISCUSSION

Standard EPA (Environmental Protection Agency) water quality microbes have been identified. Many of these microbes are common in the environment, but should not be found in potable water, and the microbes are often used as a measure of water purity. Some of these microbes are also found elsewhere, many are associated with animal waste and pollution. Some are pathogenic.

The national average for viruses ranges from less than 1 to 65 unique peptides. Five groups are discussed in detail because they stand out from the field, with three having more than 50 unique peptides and two having more than 20.

The classification and addition of EPA water microbes are ongoing and some of them may be under-speciated at the moment because many microbes are not sequenced and remain un-named. Water viruses have a high mutation rate, and the number of species or strains within a genus and species can rapidly change. Likewise, additional strains of bacteria are constantly being updated. The rapid increase in sequencing will contribute to future identification and cement phylogenic relationships. Nevertheless, in this chapter water microbes were detected and identified from honeybees. Some of these microbes may have been collected from foraging others may have been from other sources.

Environmentalists are encouraged to study these results and to ascertain the mutual relationships and assess these relationships within a region, area or a locality.

The number of microbes that can be identified is continuing to increase due to the increasing number of sequences added to the National Center for Biotechnology Information. More than 500,000 microbes have been sequenced; the number increases daily.

The ability to add new microbes has been discussed in Chapter 2, but it remains important to remember that as new microbes are sequenced and added to the water data group, the samples that have been analyzed can be re-analyzed using the old computer files. In this manner there is no need to collect a new sample, and since large archived sample sets can be re-examined and a search can be made for new microbes, considerable time is saved and an historical record can be developed. An example of this is demonstrated in Chapter 5 where the SARS and COVID viruses were downloaded into their own data group. Recently over 450 new viruses were added to the virus data group.

Finally, considering all the microbe species that exist everywhere and are being discovered, it is possible to consider that we live in a very rich environment of microbes of which viruses, bacteria and fungi are a large part. There is so much potential for the type of information collected by the honeybees. In their regular work, honeybees collect all the microbe information about their working environment. They carry the microbes with them; all we have to do is sample, detect and identify. Microbes of all sorts are found with the honeybees, and the diversity of these microbes may be useful in measuring or monitoring changes to the environment.

11 Discussion

This chapter discusses the detection and identification of microbes collected by honeybees. Several hundreds of bacteria, fungi and viruses have been collected from honeybees from all over the United States.

Results have been used to determine a national average and regional averages for California, Florida, Idaho, Iowa and Montana. The regional averages have been compared to the national average. Many questions arise from this collection of data. The results show that there are regional differences with some regions above the national average and some below. In some cases, groups of microbes stand out.

How can this information be used is discussed as well as recommendations for further work, including the collection of this type of data for seasons to determine a seasonal average, seasonal averages for regions and more. Questions such as why are the regions different should be answered as well as why are some particular microbes in such high numbers.

It is interesting to see that COVID-19 was detected and identified in abundance only in one region, only a small detection in others and still no detection at all in two regions. Perhaps the honeybee can be used to monitor the movement of microbes and help track epidemics such as COVID-19.

11.1 WHAT TO MAKE OF ALL THE MICROBES CARRIED BY HONEYBEES?

Technology and especially the ability to detect and identify microbes in the honeybees have been on a fast pace for several years. In the 1990s new methods such as polymerase chain reaction (PCR) and Integrated Virus Detection System (IVDS) were invented and developed, computers became faster and software to analyze results from mass spectrometers was invented and developed. These advances have led to a plethora of data. These advancements are dramatically illustrated by the rise in the number of sequenced microbes starting with none, to a few, to a few hundred to over 200,000 today. Future advancements can be expected to result in hand held instruments using this approach.

One result of present advancements is the ease by which we can examine microbes associated with honeybees. For that matter, microbes associated with nearly anything. The focus on the honeybee collection of microbes has shown in this short text the diversity of the collection and the extent of the collection, not to mention the unexpected discoveries. Remember that the current sequenced list represents only a very small fraction of the potential total. We can expect to see additional sequences added moving forward. These can be added easily to the existing data groups, and the files analyzed in this book can be re-analyzed with the new listings. It will be useful to see more eukaryotes, currently at 11,965, prokaryotes, currently listing 251,975 and viruses which currently list 38,431. I believe that patterns will emerge which show both the occurrence of these organisms and their concentrations over time measured by days, weeks, months, seasons and years. These patterns will emerge as we examine large numbers of samples and will be useful in many fields of study.

In this book we have looked at those microbes carried back to the hive by honeybees. The samples are from hives and are from commercial bees in general. The bees are being used to pollinate crops, orchards and other related activities and produce honey. Recall the nature of the bee, they find a source of nectar or pollen and then go straight out and get it just like a long-haul trucking service. They tend to be very focused on their collection mission. This is somewhat different from the urban bees which are not looking at acres of nectar and pollen but usually a smaller source. Commercial bees are taken to a plant source, urban bees need to search and find them. As a result, it could be expected that those areas that commercial honeybees frequent would have a different mix of microbes than urban bees. This difference was beyond the scope of this effort and would certainly be interesting material for a future book on the subject.

Let us examine the findings among honeybees. Each chapter illustrated the collection of multiple microbes collected by the honeybees. From those microbes associated with the bee gut to common bacteria and viruses it was interesting to see from the two levels of detection: low levels such as that seen among the bacteria (forensic) and high levels as seen in *Nosema* and the fungi and viruses (clinical) that the honeybees collect a rich and varied number of genomic information representative of the environment in which they are operating.

11.1.1 What Is Common from All the Sources?

We will call this the "discovered biome" as it contains numerous bacteria, viruses, fungi and plants discovered during analysis. This biome contains some similarities when all sources are analyzed. Let us consider the clinical detection ones first. There are several microbes that appear among all the honeybee samples. What can be said about these are that they tend to be found in many places and may represent a natural distribution and natural reservoir of microbes.

Microbes showing up at the more sensitive or forensic level of detection tell a different story. These microbes may be just part of the biological cloud that surrounds us and these microbes may be just floating around and represent a normal biological flora of the natural biome that the honeybees operate. It would be interesting to examine the vertical distribution of this flora and see how high it reaches in the atmosphere. This could be expected to vary according to time of day and time of the year with predicable distributions seen by others in limited studies. This biome could be expected to be different from area to area and in particular urban areas when compared with agricultural areas or even between geological areas. The honeybees could be the collection method and the mass spectrometry proteomics (MSP) with the ABOid software could be the analysis method. The results would certainly be interesting and useful.

11.1.2 What about Monitoring Water Quality?

Chapter 10 contains the results of analysis of honeybee samples from a water quality point of view. It was possible to create a data group that contains microbes of interest by the EPA (Environmental Protection Agency) in regard to water quality. This is the list that the EPA would like to use to test drinking water and in particular commercial water supplies.

The honeybees need water and they will stop and collect water in the course of their activity. They appear to not be particularly selective on the water source, but I would suspect that it is usually some water supply nearby. It was easy enough to analyze the honeybee samples using the water data group. The results were interesting. Many of the microbes in the data group were routinely detected and identified at both the clinical and the forensic levels. The water quality of an area where the honeybees are operating could be monitored using this means of analysis. The importance of this could be realized when you consider the impact of flooding that may occur from time to time and in particular in the aftermath of a major storm or hurricane. The bees would forage, and the analysis of the water could be useful in determining the presence of undesirable microbes, water safety and quality for consumption and the need for treatment and remediation.

Temporal measurement could follow the occurrence of microbes of interest and be used to monitor water quality in local water sources. It would be interesting to determine how the microbes vary during the summer and winter, for example. This can be refined to see what the effect of rainfall and other environmental factors may be on the local water sources.

Regarding just the honeybee, it would seem that knowing the water quality that the honeybees are using would be an important factor in bee health. It was beyond this effort, but the change in water quality between pollination sites could be an important factor for the health of commercial honeybees. Thus, the normal water quality could be compared with the water quality of different commercial sites to determine if there is a correlation with honeybee health.

11.2 CHANGES TO THE NATURAL BIOME, ADVANCED WARNING AND GENERAL CONSIDERATIONS

The natural biome moves around with the weather, and it would not be surprising to find African microbes carried across the Atlantic by the wind or similar findings for other major weather systems. The normal or natural biome can be expected to be stirred up by natural events such as thunderstorms, rain, tornados, hurricanes and other similar activities, natural or manmade. The results of such disturbances could be expected to change the local microflora, for example, flooding, damage to infrastructure, and movements of populations. All of these disturbances to the natural biome and the perchance encounter with a pathogen indicate the need for an early warning capability to ascertain a change in the national average to the natural biome and determine if it has changed into one more sinister to a population. Knowing this information can guide remediation and safety needs.

11.2.1 Seasonal Changes and the Environment

Charts that need to be made are those that cover a period of time as measured in days, weeks, months, seasons and years. The diversity of microbes collected by the honeybee is indicative of very proactive means to monitor the environment. This diversity evidently changes in regard to region and time of the year. When the commercial honeybees are moved from one pollination site to another, this diversity changes. I would suggest that these studies need to be made.

An important criterion for monitoring the environment for microbes is the ability to not be limited in which microbes you are seeking. The interaction among the trillions of microbes can be just as important as the detection of a single virus, for example: the detection of obscure bacteriophages can indicate the presence of a bacteria, and likewise the detection of a plant virus can be an indirect detection of a plant. Indeed, the forensic evaluation of the microbes in the environment can be indicators of greater events taking place. The SARS-CoV group is seen naturally. There are over 100 different coronaviruses that occur in the environment.

11.2.2 Advance Warning of Infectious Disease

The COVID-19 outbreak occurred during the writing of this manuscript. It was a simple enough effort to download the SARS sequences and build a SARS data group as discussed in Chapter 5. It could be expected that these honeybees would not come close to finding these viruses, but the fact that they did may be useful in the detection and early warning of a viral or indeed a microbial outbreak. To bring up the expected difference between urban and commercial honeybee samples, it could be expected that urban samples would contain a higher number of detections and identifications for microbes of interest, especially SARS and influenza.

Routine monitoring of general honeybee activities could provide the early warning or indeed an indication of the end of a particular contagion.

11.2.3 Bacteria, Viruses and the Plants

Among the discoveries were viruses that are causative agents of diseases in plants. These viruses give an indirect detection of the associated plant. For example, tomato leaf virus, which causes a leaf disease in tomato plants. It is an important commercial problem for those growing tomatoes. Likewise, tobacco mosaic virus and others. The plants which are among the eukaryotes have only 1,340 sequences and only 3 completely which are algae. This limits the detection of pollen and nectar sources and other plants of environmental importance. It is expected with the recent advances in science that we will see more and more plants joining the fully sequenced list, and then they can be added to the current data groups.

Bacteria phages associated with particular bacteria have been detected. These give evidence of the presence of those bacteria that have not been sequenced. This capability provides another method to detect, at least indirectly, a new microbe.

Likewise, when 460 new viruses were added to the virus data group, many new phages were added and several new viruses. The RAW files were simply re-examined, and the new viruses were detected if present.

11.3 FOLLOWING MICROBES OVER A TEMPORAL PERIOD

Environmentalists have often wondered on how to measure the effects of different activities on the environment. It appears that the honeybee may provide an interesting opportunity to monitor the environmental biome. This biome could be expected to change as the environment changes and react to changes as they are being made. Following seasonal changes could provide the basic biome levels of microbes and can be used as the basis for determining what is normal and what is change. The graphs and charts in this process would be useful information to understanding the relationship between our environment and our activities.

11.4 GENERAL CONSIDERATIONS

Just because microbes have been detected and identified does not indicate there is an infectious or zoonotic threat that is to be determined by others. The wide diversity of microbes detected and identified do show that we do not live in a sterile environment. Microbes are widely dispersed and their numbers vary by regions. The numerous microbes among the bacteria, fungi and viruses are illuminating when we consider the microflora cloud. Microbes are both numerous and diverse.

Regarding the application of the MSP/IVDS/ABOid system. We have discovered that honeybees collect nearly every microbe in their environment and specific encounters with urban setting. Consider, for example, the case of a demonstration hive in the window of a science building. Analyzing the honeybees from that demonstration hive revealed that all the microbes being studied in the science building were detected and identified. Honeybees could be a unique collector for different areas of interest – urban, rural, industrial and around buildings of special interest such as hospitals, science laboratories and buildings with unknown utility. This capability has already been demonstrated.

MSP/ABOid has been used to analyze thousands of complex environmental samples over many years and is being used to detect and identify microbes in honeybees. In this capacity, MSP/ABOid has identified hundreds of microbes including bacteria, viruses and fungi that make up a common microbial flora, in addition to those microbes that are of special interest to the honeybees. Some of these honeybee microbes include those that exist in the honeybee gut and are important in processing glucose and performing other vital functions. Likewise, microbes that impact honeybee health such as sacbrood, deformed wing and paralysis viruses can be detected and identified. MSP/ABOid and especially the resulting computer files that are archived allow these microbes to be followed over the seasons and regions. The ability to add new microbes and re-analyze the computer files has allowed current and updated comparisons to be made for daily, seasonal and yearly samples.

Given this experience and the success of MSP/ABOid, it can easily be seen how MSP/ABOid is important to the national interests. The MSP/ABOid system can be located forward, close to the first sources of a threat microbe. MSP results produce a computer file which can be sent via Internet to a central site where ABOid analysis can be performed to standardize results and provide a quick data analysis. Analyzed data can then be used to brief those in authority. Alert of new microbes (threats) in this manner would be quick and provide timely data to illustrate the occurrence and spread of a microbe. Since it is important, if not vital, that any method for detecting and identifying microbes not to be limited (number of microbes that can be detected and identified) in detecting and identifying all new and emerging threats, MSP/ABOid could provide this frontline early warning. When combined with the IVDS system, both rapid early warning (IVDS), and identification of microbes (MSP/ABOid) is achieved. Since the honeybee is located around the world, they could serve as the bio-collector for the detection and identification means.

It is a feature of both the MSP/ABOid and IVDS instruments in that following utilization they can easily be placed in standby or storage mode until needed again. This is possible because there are no sensitive reagents and no "use-by-date". It is our experience that once an IVDS/MSP/ABOid unit is set up and operating that nearly anyone with an aptitude can be trained to run and use the equipment.

We downloaded the sequences for the EPA list of microbes of interest in waste water monitoring and quality control. Since water is important to honeybees, we were interested in the results of this analysis. We have run hundreds of honeybee samples through this water data group with great success. In this manner we have also demonstrated the IVDS/MSP/ABOid instrumentation to monitor, detect and identify microbes of interest in water.

Given the success of IVDS/MSP/ABOid in monitoring and classifying the microbes in complex environmental situations and the ease of sampling individuals, it looks compelling to utilize the IVDS/MSP/ABOid instrumentation anywhere there is a need for microbe detection that includes bacteria, viruses and fungi. All this is accomplished by collecting and analyzing one sample, one effort, low cost with an archival record and the ability to re-analyze when microbe updates are available.

We have discovered a practical assistant in monitoring the environment, the honeybee. The honeybee and related species can be found nearly everywhere and can be imported to sites of interest. We have analyzed thousands of samples from honeybees from all over the United States and some from different parts of the world. The enormous number of microbes detected and identified in these samples tells a story for each of the regions where they were collected and illustrates the impact of seasonal and environmental conditions. Urban honeybees have a different diversity than rural honeybees, and both are different from commercial honeybees. This insect could provide a means to sample large areas to a given measure of the microbial loading. It can be imagined that a honeybee hive on top of a building could provide an insight into the microbial conditions surrounding that site. Although honeybees do not fly year around, other collectors could be used to collect samples during this time.

The SARS-CoV-2 (COVID-19) virus can be simply detected and quantified by the IVDSand identified by MSP/ABOid. Since the IVDS system is a particle counter that can count virus-sized particles, it has become a quick and reliable method to screen for viruses (minutes). IVDS has analyzed several thousand complex environmental samples and currently is being used to detect and identify viruses in honeybees. Likewise, MSP/ABOid has identified hundreds of microbes including bacteria, viruses and fungi with genomic accuracy.

A valued feature of the IVDS/MSP/ABOid system is that new strains of the SARS-CoV-2 can be detected by IVDS without any new adjustments. New strains or indeed any other microbe can be added to the MSP/ABOid data groups as a sequence for a new microbe becomes available. The addition of a new sequence takes a couple of hours, and the samples can be re-analyzed using the MSP data file. This is important for a couple of reasons. First, new microbes can quickly be added to the existing data groups, and second, analysis for the new microbe can begin immediately. Since old files can be re-analyzed, a search for the new virus from old files can help find the origin or origins of the new virus from the historical record.

The MSP/ABOid monitoring and analysis of the environment could provide timely intelligence on emerging disease and microbial threats. With the ability to sample, analyze and process data quickly, this system is indicated for frontline use. It can be networked (Internet) so that areas separated by large distances can compare results and operate almost together in detecting a geographical threat. It can further be used to monitor the movement and spread of an infectious threat and provide timely and reliable information to decision makers.

This book is only a start on reporting microbial diversity in honeybees. The addition of new viruses and the analysis of other regions within the United States and indeed the world will provide new and continuing information on microbial diversity as reported by the honeybees.

Bibliography

Alderwick, L. J., Harrison, J., Lloyd, G. S., & Birch, H. L. (2015). The mycobacterial cell wall –Peptidoglycan and arabinogalactan. *Cold Spring Harbor Perspectives in Medicine*, 5 (8), a021113.

Alippi, A. M., & Reynaldi, F. J. (2006). Inhibition of the growth of *Paenibacillus larvae*, the causal agent of American foulbrood of honeybees, by selected strains of aerobic spore-forming bacteria isolated from apiarian sources. *Journal of Invertebrate Pathology*, 91 (3), 141–146.

Allepuz, A., Casal, J., & Beltrán-Alcrudo, D. (2019). Spatial analysis of lumpy skin disease in Eurasia – Predicting areas at risk for further spread within the region. *Transboundary and Emerging Diseases*, 66 (2), 813–822.

Arihara, K. O. (1998). *Lactobacillus acidophilus* group lactic acid bacteria applied to meat fermentation. *Journal of Food Science*, 63 (3), 544–547.

Balaban, M., & Hendrixson, D. R. (2011). Polar flagellar biosynthesis and a regulator of flagellar number influence spatial parameters of cell division in Campylobacter jejuni. *PLOS Pathogens*, article e1002420. https://doi.org/10.1371/journal.ppat.1002420

Balamurugan, V. (March 2009). A polymerase Chain reaction strategy for the diagnosis of Camelpox. *Journal of Veterinary Diagnostic Investigation*, 21 (2), 231–237.

Blaser, M. (1997). Epidemiologic and clinical features of *Campylobacter jejuni* infections. *The Journal of Infectious Diseases*, 176, S103–S105.

Bock, E., Sundermeyer-Klinger, H., & Stackebrandt. (1983). New facultative lithoautotrophic nitrite-oxidizing bacteria. *Archives of Microbiology*, 136, 281–284.

Breman, J. G. (2000). Monkeypox: an emerging infection for humans? In S. W. Craig and J. Hughes (Eds.), Emerging Infections 4 (pp. 45–67). Washington, DC: ASM Press.

Bromenshenk, J., Henderson, C. B., Wick, C. H., Stanford, M. F., Zulich, A. W., Jabbour, R. E., … Cramer Jr, R. A. (2010). Iridovirus and microsporidian linked to honey bee colony decline. *PLOS ONE*, 5 (10), e13181.

Chase-Topping, M., Gally, D., Low, C., Matthews, L., & Woolhouse, M. (2008). Super-shedding and the link between human infection and livestock carriage of *Escherichia coli* O157. *Nature Reviews Microbiology*, 6, 904–912.

Dixon L. K., Abrams, C. C., Chapman, D. G., & Zhang, F. (2008). African swine fever virus. In Thomas C. Mettenleiter and Francisco Sobrino (Eds.), *Animal viruses: Molecular biology*. Madrid, Spain: Caister Academic Press.

Dodds, W. K., Gudder, D. A., & Mollenhauer, D. (1995). The ecology of *Nostoc*. *Journal of Phycology*, 31 (1), 2–18.

DOE/Joint Genome Institute. (2019). *Bee gut microbes have a division of labor when it comes to metabolizing complex polysaccharides*. Retrieved from phys.org.

El-Niweiri, M. A., Moritz, R. F., & Lattorff, H. M. (2019). The invasion of the dwarf honeybee, *Apis florea*, along the river Nile in Sudan. *Insects*, 10, 10–11.

Euzeby, J. (1997, August 23). List of bacterial names with standing in Nomenclature: A folder available on the Internet. *International Journal of Systematic Bacteriology*, 590–592. Retrieved from LPSN: https://www.bacterio.net/phylum#acidobacteria

Fairbrother, J. (1999). *Escherichia coli* infections in farm animals. In J. L. Howard and R. Smith (Eds.), *Current Veterinary Therapy: Food Animal Practice* (pp. 328–330). Philadelphia, USA: Saunders Company.

Ferlay J., Colombet, M., Soerjomataram, I., Mathers, C., Parkin, D. M., Piñeros, M., … Bray, F. (2018). Estimating the global cancer incidence and mortality in 2018: GLOBOCAN sources and methods. *International Journal of Cancer*, 44 (8), 1941–1953.

Foodsafety.gov. (2020, August 10). *Campylobacter*. Retrieved from www.foodsafety.gov

Graham, S. (2003). Bacterial battery converts sugar into electricity. *Scientific American*. Retrieved from https://www.scientificamerican.com/article/bacterial-battery-convert/

Gram, H. (1884). Über die isolierte Färbung der Schizomyceten in Schnitt- und Trockenpräparaten. *MMW – Fortschritte der Medizin*, 2, 185–189.

Guez, D., Subias, L., & Griffin, A. S. (2017). Colour and shape preferences of *Apis cerana* (Java genotype) in Australia. *Bulletin of Insectology*, 70, 267–272.

Hsuan, H. M., Salleh, B., & Zakaria, L. (2011). Molecular identification of *Fusarium* species in *Gibberella fujikuroi* species complex from rice, sugarcane and maize from Peninsular Malaysia. *Journal of Molecular Sciences*, 12 (10), 6722–6732.

Hugenholtz, P. (2002). Exploring prokaryotic diversity in the genomic era. *Genome Biology*, 3 (2).

Imhoff, J. F. (2006). The phototrophic β-Proteobacteria. *In the prokaryotes* (pp. 593–601). New York: Springer.

Jabbour, R. E., Wade, M. M., Deshpande, S. V., Stanford, M. F., Wick, C. H., Zulich, A. W., & Snyder, A. P. (2010). Identification of *Yersinia pestis* and *Escherichia coli* strains by whole cell and outer membrane protein extracts with mass spectrometry-based proteomics. *Journal of Proteome Research*, 9 (7), 3647–3655.

Jeon, S. J., Elzo, M., DiLorenzo, N. Lamb, G. C., & Jeong, K. C. (2013). Valuation of animal genetic and physiological factors that affect the prevalence of *Escherichia coli* O157 in cattle. *PLOS ONE*, 8 (2), e55728.

Karl, A. D., Michaels, A., Bergman, B., Capone, D., Carpenter, E., Letelier, R., ..., & Stal, L. (2002). Dinitrogen fixation in the world's oceans. *Biogeochemistry*, 47–98.

Kešnerová, L., Mars, R. A., Ellegaard, M. K. L., Troilo, M., Sauer, U., & Engel, P. (2017). Disentangling metabolic functions of bacteria in the honey bee gut. *PLOS Biology*, 15 (12).

Kešnerová, L., Moritz, R., & Engel, P. (2016). *Bartonella apis* sp. nov., a honey bee gut symbiont of the class Alphaproteobacteria. *International Journal of Systematic and Evolutionary Microbiology*, 66, 414–421.

Khodakevich, L., Ježek, Z., & Kinzanzka, K. (1986). Isolation of monkeypox virus from wild squirrel infected in nature. *Lancet*, 1 (8472), 98–99.

Kirk, P. M., Cannon, P., Minter, D., & Stalpers, J. (2008). *Dictionary of the fungi* (10th ed.). Wallingford, UK: CAB International.

Kluge, J., Beran, G. H. H., & Platt, K. (1999). Pseudorabies (Aujeszky's disease). In B. L. Straw, S. O'Altaire, W. L. Mengeling and D. J. Taylor (Eds.), *Diseases of swine*. 8th Edition (pp. 233–246). Ames, USA: Iowa State University Press.

Kwong, W. A. & Moran, N. A. (2013). Cultivation and characterization of the gut symbionts of honey bees and bumble bees: Description of *Snodgrassella alvi* gen. nov., sp. nov., a member of the family Neisseriaceae of the Betaproteobacteria, and *Gilliamella apicola* gen. nov., sp. nov., a member of Orbaceae fam. nov., Orbales ord. nov., a sister taxon to the order "Enterobacteriales" of the Gammaproteobacteria. *International Journal of Systematic and Evolutionary Microbiology*, 63 (6), 2008–2018.

Lee, H., Chun, J., Moon, E., Ko, S., Lee, D., Lee, H., & Bae, K. (2001). Hahella chejuensis gen. nov., sp. nov., an extracellular-polysaccharide-producing marine bacterium. *International Journal of Systematic and Evolutionary Microbiology*, 51 (2), 661–666.

Lichtinger T., Reiss, G., & Benz, R. (2000). Biochemical identification and biophysical characterization of a channel-forming protein from *Rhodococcus erythropolis*. *Journal of Bacteriology*, 182 (3), 764–770.

Machado, G., Korennoy, F., Alvarez, J., Picasso-Risso, C., Perez, A., & VanderWaal, K. (2019). Mapping changes in the spatiotemporal distribution of lumpy skin disease virus. *Transboundary and Emerging Diseases*, 66 (5), 2045–2057.

Maphosa, F., Lieten, S. H., Dinkla, I., Stams, A. J., Smidt, H., & Fennell, D. E. (2012). Ecogenomics of microbial communities in bioremediation of chlorinated contaminated sites. *Frontiers in Microbiology*, 3, Article 351.

Miyamoto. (2006). *Lactobacillus harbinensis*. Retrieved from https://www.namesforlife.com/10.1601/nm.9813

Moran, N. H., Hansen, A. K., Powell, J. E., & Sabree, Z. L. (2012). Distinctive gut microbiota of honey bees assessed using deep sampling from individual worker bees. *PLOS ONE*, 7 (4), e36393.

Moss, B. (2007). Poxviridae: The viruses and their replication. In D. M. Knipe, P. M. Howley, D. E. Griffin, R. A. Lamb, M. A. Martin, B. Roizman and S. E. Straus (Eds.), *Fields virology*. (5th ed., pp. 2905–2946). Philadelphia, PA: Lippincott, Williams & Wilkins.

Moss, B. (2013). Poxviridae. In D. M. Knipe and P. M. Howley (Eds.), *Fields virology* (6th ed., pp. 2129–2159). Philadelphia, PA: Lippincott Williams & Wilkins.

Peterson, S., Ito, Y., Horn, B., & Goto, T. (2001). *Aspergillus bombycis*, a new aflatoxigenic species and genetic variation in its sibling species, *A. nomius*. *Mycologia*, 93 (4), 689–703.

Poddar, S., & Khurana, S. (2011). Geobacter: The electric microbe! Efficient microbial fuel cells to generate clean, cheap electricity. *Indian Journal of Microbiology*, 51 (2), 240–241.

Powell, J. E., Leonard, S. P., Kwong, W. K., Engel, P., & Moran, N. A. (2016, November 29). Genome-wide screen identifies host colonization. *PNAS*, 113 (48), 13887–13892.

Pruimboom-Bre, I. M., Morgan, T. W., Ackermann, M. R., Nystrom, E. D., Samuel, J. E., Cornick, N, A., & Moon, H. W. (2000). Cattle lack vascular receptors for *Escherichia coli* O157:H7 Shiga toxins. *Proceedings of the National Academy of Sciences of the United States of America* (pp. 10325–10329).

Raymann, K., Coon, K. L., Shaffer, Z., Salisbury, S., & Moran, N. (2018, October). Pathogenicity of *Serratia marcescens* strains in honey bees. *mBio*, e01649-18. Retrieved from mBio.

Reinhold-Hurek, B., Hurek, T., Gillis, M., Hoste, B., Vancanneyt, M., Kersters, K., & De Ley, J. (1993). Azoarcus gen. nov., nitrogen-fixing Proteobacteria associated with roots of kallar grass (*Leptochloa fusca* (L.) Kunth), and description of two species, *Azoarcus indigens* sp. nov. and *Azoarcus communis* sp. nov. *International Journal of Systematic Bacteriology*, 43 (3), 574–584.

Rheinbaden, F., Gebel, J., Exner, M., & Schmidt, A. (2007). Environmental resistance, disinfection, and sterilization of poxviruses. In A. A. Mercer, A. Schmidt and O. Weber (Eds), *Poxviruses* (pp. 397–405). Basel, Switzerland: Birkhauser Verlag.

Roossinck, M. (2016). *Virus: An illustrated guide to 101 incredible microbes*. Princeton University Press.

Ruarn K. J., R. C. George, & Ryan, K. J. (2004). *Sherris medical microbiology* (4th Ed.). McGraw Hill.

Saberioon, M. M., Mardan, M., Nordin, L., Alias, M. S., & Gholizadeh, A. (2010). Predict location(s) of *Apis dorsata* nesting sites using remote sensing and geographic

information system in Melaleuca forest. *American Journal of Applied Sciences*, 7 (2), 252–259.
Scardovi, V., & Trovatelli, L. (1969). New species of bifid bacteria from *Apis mellifica* L. and *Apis indica* F.: A contribution to the taxonomy and biochemistry of the genus *Bifidobacterium. Zentralblatt für Bakteriologie, Mikrobiologie und Hygiene II*, 123 (1), 64–88.
Schaechter, M. I. (2006). Prokaryotic microbes: Phylum B10: Cyanobacteria. *Microbe*, Chapter 15, pp. 306–307. Washington DC: AMS Press.
Selle, K. M., Klaenhammer, T. R., & Russell, W. M. (2014). LACTOBACILLUS | *Lactobacillus acidophilus*. In C. A. Batt and M.-L. Tortorello (Eds.), *Encyclopedia of food microbiology* (2nd ed., pp. 412–417). Academic Press.
Sergeev, A. A., Kabanov, A. S., Bulychev, L. E., Sergeev, A. A., Pyankov, O. V., Bodnev, S. A., … Sergeev, A. N. (2017). Using the ground squirrel (*Marmota bobak*) as an animal model to assess monkeypox drug efficacy. *Transboundary and Emerging Diseases*, 64 (1), 226–236.
Shchelkunov, S. S., Marennikova, S. S., Moyer, R. W. (2005). *Orthopoxviruses pathogenic for humans*. Berlin: Springer-Verlag.
Silimela, M., & Korsten, L. (2007). Evaluation of pre-harvest Bacillus licheniformis sprays to control mango fruit diseases. *Crop Protection*, 26 (10), 1474–1481.
Singh, A. B., & Dahiya, P. (2008). Aerobiological researches on pollen and fungi in India during the last fifty years: An overview. *Indian Journal of Allergy, Asthma & Immunology*, 22, 27–38.
Singh, R. K., Balamurugan, V., Bhanuprakash, V., Venkatesan, G., & Hosamani, M. (2012). Emergence and reemergence of vaccinia-like viruses: Global scenario and perspectives. *Indian Journal of Virology*, 22, 1–11.
Snydman, D. R., Jacobus, N. V., McDermott, L. A., Golan, Y., Hecht, D. W., Goldstein, E. J., … Rosenbla. (2010). Lessons learned from the anaerobe survey: Historical perspective and review of the most recent data (2005–2007). *Clinical Infectious Diseases*, 50, S26–S33.
Sudhagar, S., & Reddy, R. (2017). Influence of elevation in structuring the gut bacterial communities of *Apis cerana* Fab. *Journal of Entomology and Zoology*, 5 (3), 434–440.
Suwannapong, G., Maksong, S., Yemor, T., Junsuri, N., & Benbow, M. E. (2013). Three species of native Thai honey bees exploit overlapping pollen resources: Identification of bee flora from pollen loads and midguts from *Apis cerana, A. dorsata* and *A. florea. Journal of Apicultural Research*, 52 (5), 196–201.
Toole, D., & Li, H. (2014). The pathology of malignant catarrhal fever, with an emphasis on ovine herpesvirus 2. *Veterinary Pathology*, 51 (2), 437–452.
Tulman, E. R., Afonso, C. L., Lu, Z., Zsak, L., Kutish, G. F., & Rock, D. L. (2001). Genome of lumpy skin disease virus. *Journal of Virology*, 75 (15), 7122–7130.
Urdaci, M., Bressollier, P., & Pinchuk, I. (2004). Bacillus clausii probiotic strains: Antimicrobial and immunomodulatory activities. *Journal of Clinical Gastroenterology*, 38, 586–590.
Vilain, S. L., Yun Luo, Hildreth, M. B., & Brözel, V. S. (2006). Analysis of the life cycle of the soil saprophyte *Bacillus cereus* in liquid soil extract and in soil. *Applied Environmental Microbiology*, 72 (7), 4970–4977.
Vodovar, N., Vallenet, D., Cruveiller, S., Rouy, Z., Barbe, V., Acosta, C., … Frédéri. (2006). Complete genome sequence of the entomopathogenic and metabolically versatile soil bacterium *Pseudomonas entomophila. Nature Biotechnology*, 24 (6), 673–679.
Wick, C. H. (2010). *Patent No. 7,850,908 B1*. United States of America.
Wick, C. H. (2013). *Patent No. 8,412,464 B1*. United States of America.
Wick, C. H. (2014). *Identifying microbes by mass spectrometry proteomics*. CRC Press.

Wick, C. H. (2015). *Integrated virus detection*. Boca Raton: CRC Press.

Wick, C. H., Standford, M. F., Zulich, A. W., Deshpande, S. V., Jabbour, R. E., Dworzanski, J. P, & McCubbin, P. E, (2012). *Patent No. 8,224,581 B1*. United States of America.

Wick, C. H., Stanford, M., Jabbour, R., Deshpande, S., McCubbin, P., Skowronski, E., & Zulich, A. (2009). Pandemic (H1N1) 2009 cluster analysis: A preliminary assessment. *Natureprecedings*.

World Animal Health Information Database (WAHID) Interface. (2020). Retrieved from http://www.oie.int/wahis/public.php?page=home

Yamaoka, Y. (2008). Helicobacter pylori*: Molecular genetics and cellular biology*. Caister: Academic Press.

Zheng, H., Perreau, J., Powell, J. E., Han, B., Zhang, Z., Kwong, W. K., … Moran, N. A. (2019). Division of labor in honey bee gut microbiota for plant polysaccharide digestion. *Proceedings of the National Academy of Sciences* (pp. 25909–25916).

Appendix A – National Average for Bacteria

Listing of bacteria detected and identified along with average number of unique peptides.

Known as:	National Bacteria Average
Acidobacteria bacterium	1.9
Acinetobacter	1.3
Aeropyrum pernix	0.8
Agrobacterium tumefaciens	3.2
Alcanivorax borkumensis	1.4
Alkalilimnicola ehrlichei	1.7
Anabaena variabilis	2.4
Anaeromyxobacter dehalogenans	2.6
Anaplasma marginale	0.4
Anaplasma phagocytophilum	0.6
Aquifex aeolicus	0.8
Archaeoglobus fulgidus	1.7
Azoarcus	2.9
Bacillus anthracis	2.3
Bacillus anthracis	2.5
Bacillus anthracis	2.5
Bacillus cereus	2.7
Bacillus cereus	2.3
Bacillus cereus	2.4
Bacillus clausii	2.0
Bacillus halodurans	1.4
Bacillus licheniformis	2.0
Bacillus subtilis	2.0
Bacillus thuringensis serovar konkukian	2.6
Bacillus thuringiensis	2.3
Bacteroides fragilis	2.0
Bacteroides fragilis	2.0
Bacteroides thetaiotaomicron	3.4
Bartonella henselae	0.7

(*Continued*)

Known as:	National Bacteria Average
Bartonella quintana	0.4
Baumannia cicadellinicola	0.3
Bdellovibrio bacteriovorus	2.1
Bifidobacterium longum	0.9
Bordetella bronchiseptica	2.2
Bordetella parapertussis	1.8
Bordetella pertussis	1.6
Borrelia afzelii	0.8
Borrelia burgdorferi	0.8
Borrelia garinii	0.8
Bradyrhizobium japonicum	4.2
Brucella abortus biovar 1	2.0
Brucella melitensis	2.0
Brucella melitensis biovar Abortus	2.1
Brucella suis	2.1
Buchnera aphidicola	0.2
Buchnera aphidicola	0.8
Burkholderia cenocepacia	3.0
Burkholderia mallei	2.8
Burkholderia pseudomallei	4.6
Burkholderia pseudomallei	3.1
Burkholderia sp.	3.2
Burkholderia thailandensis	3.4
Burkholderia xenovorans	4.1
Campylobacter jejuni	0.5
Campylobacter jejuni	0.6
Candidatus Blochmannia floridanus	0.2
Candidatus Blochmannia pennsylvanicus str.	0.2
Candidatus Pelagibacter ubique	0.9
Candidatus Protochlamydia amoebophila	0.9
Carboxydothermus hydrogenoformans	1.0
Caulobacter crescentus	2.0
Chlamydia muridarum	0.4
Chlamydophila abortus	0.4
Chlamydophila felis	0.7
Chlamydophila pneumoniae	0.4
Chlamydophila pneumoniae	0.4
Chlamydophila pneumoniae	0.3
Chlamydophila pneumoniae	0.4
Chlorobium chlorochromatii	0.9
Chlorobium tepidum	1.0
Chromobacterium violaceum	2.2

(*Continued*)

Known as:	National Bacteria Average
Chromohalobacter salexigens	1.5
Clostridium acetobutylicum	1.7
Clostridium perfringens	1.1
Clostridium perfringens	1.3
Clostridium perfringens	1.6
Clostridium tetani	1.0
Colwellia psychrerythraea	1.6
Corynebacterium diphtheriae	0.8
Corynebacterium efficiens	1.6
Corynebacterium glutamicum	1.7
Corynebacterium jeikeium	1.0
Coxiella burnetii	0.8
Cytophaga hutchinsonii	1.7
Dechloromonas aromatica	2.2
Dehalococcoides ethenogenes	3.2
Dehalococcoides sp.	0.8
Deinococcus geothermalis	1.4
Deinococcus radiodurans	1.3
Desulfitobacterium hafniense	2.1
Desulfotalea psychrophila	1.3
Desulfovibrio desulfuricans	1.4
Desulfovibrio vulgaris	2.2
Ehrlichia canis str.	0.4
Ehrlichia chaffeensis str.	0.5
Ehrlichia ruminantium	0.4
Ehrlichia ruminantium	0.4
Enterococcus faecalis	1.5
Erwinia carotovoa	1.9
Erythrobacter litoralis	1.5
Escherichia coli	1.8
Escherichia coli	1.8
Escherichia coli	2.3
Escherichia coli	1.6
Escherichia coli	2.0
Escherichia coli	2.0
Escherichia coli	2.1
Escherichia coli	1.6
Fowlpox Virus	0.2
Francisella tularensis	1.2
Francisella tularensis	1.2
Francisella tularensis	1.1
Francisella tularensis	1.2

(*Continued*)

Known as:	National Bacteria Average
Frankia alni	4.6
Frankia sp.	2.4
Fusobacterium nucleatum	1.2
Geobacillus kausophilus	1.9
Geobacter metallireducens	1.9
Geobacter sulfurreducens	2.0
Gloeobacter violaceus	1.7
Gluconobacter oxydans	1.6
Granulibacter bethesdensis	1.3
Haemophilus ducreyi	0.7
Haemophilus influenzae	0.7
Haemophilus influenzae	0.7
Haemophilus somnus	1.0
Hahella chejuensis	3.0
Haloarcula marismortui	1.4
Halobacterium	0.8
Haloquadratum walsbyi	0.9
Helicobacter acinonychis str.	0.7
Helicobacter hepaticus	1.0
Helicobacter pylori	1.1
Helicobacter pylori	1.0
Helicobacter pylori	1.0
Hyphomonas neptunium	2.0
Idiomarina loihiensis	1.2
Jannaschia sp.	1.8
Klebsiella pneumoniae	2.0
Lactobacillus acidophilus	1.4
Lactobacillus delbrueckii	0.9
Lactobacillus johnsonii	1.1
Lactobacillus plantarum	0.9
Lactobacillus sakei	0.8
Lactobacillus salivarius	0.6
Lactococcus lactis	1.3
Lawsonia intracellularis	0.7
Legionella pneumophila	1.5
Legionella pneumophila	1.5
Legionella pneumophila	1.6
Leifsonia xyli	1.0
Leptospira interrogans serovar Copenhageni	2.2
Leptospira interrogans serovar lai	2.4
Listeria innocua	1.3
Listeria monocytogenes	1.1

(*Continued*)

Known as:	National Bacteria Average
Listeria monocytogenes	1.0
Magnetospirillum magneticum	2.2
Mannheimia succiniproducens	1.2
Maricaulis maris	1.6
Mesoplasma florum	0.6
Mesorhizobium loti	3.4
Mesorhizobium sp.	2.2
Methanocaldococcus jannaschii	1.2
Methanococcoides burtonii	1.0
Methanococcus maripaludis	0.8
Methanopyrus kandleri	0.8
Methanosarcina acetivorans	2.4
Methanosarcina barkeri str.	1.5
Methanosarcina mazei	1.8
Methanosphaera stadtmanae	1.1
Methanospirillum hungatei	1.6
Methanothermobacter thermautotrophicus	0.5
Methylobacillus flagellatus	1.6
Methylococcus capsulatus	1.4
Moorella thermoacetica	1.4
Mycobacterium avium	2.6
Mycobacterium bovis	1.9
Mycobacterium leprae	1.1
Mycobacterium sp.	2.8
Mycobacterium tuberculosis	1.6
Mycobacterium tuberculosis	1.8
Mycoplasma capricolum	0.3
Mycoplasma gallisepticum	0.5
Mycoplasma genitalium	0.2
Mycoplasma hyopneumoniae	0.3
Mycoplasma hyopneumoniae	0.4
Mycoplasma hyopneumoniae	0.4
Mycoplasma mobile	0.2
Mycoplasma mycoides	0.6
Mycoplasma penetrans	0.6
Mycoplasma pneumoniae	0.5
Mycoplasma pulmonis	0.4
Mycoplasma synoviae	0.3
Myxococcus xanthus	5.0
Nanoarchaeum equitans	0.2
Natronomonas pharaonis	1.0
Neisseria gonorrhoeae	0.8

(*Continued*)

Known as:	National Bacteria Average
Neisseria meningitidis	0.8
Neisseria meningitidis	0.8
Neorickettsia sennetsu str.	0.4
Nitrobacter hamburgensis	2.6
Nitrobacter winogradskyi	1.9
Nitrosococcus oceani	1.5
Nitrosomonas europaea	1.4
Nitrosomonas eutropha	1.3
Nitrosospira multiformis	1.2
Nocardia farcinica	3.3
Nostoc sp.	2.8
Novosphingobium aromaticivorans	2.1
Oceanobacillus iheyensis	1.4
Onion Yellows Phytoplasma	0.5
Pasteurella Multocida	0.7
Pelobacter carbinolicus	1.6
Pelodictyon luteolum	1.2
Photobacterium profundum	1.8
Photorhabdus luminescens	1.9
Picrophilus torridus	0.7
Polaromonas sp.	2.5
Porphyromonas gingivalis	1.0
Prochlorococcus marinus	1.0
Prochlorococcus marinus	1.1
Prochlorococcus marinus	0.9
Prochlorococcus marinus str.	1.0
Prochlorococcus marinus str.	1.2
Propionibacterium acnes	1.0
Pseudoalteromonas atlantica	1.8
Pseudoalteromonas haloplanktis	1.7
Pseudomonas aeruginosa	3.4
Pseudomonas entomophila	2.5
Pseudomonas fluorescens	2.7
Pseudomonas fluorescens	2.8
Pseudomonas putida	2.9
Pseudomonas syringae pv. phaseolicola	2.0
Pseudomonas syringae pv. syringae	2.1
Pseudomonas syringae pv. tomato	2.5
Psychrobacter arcticus	0.8
Psychrobacter cryohalolentis	1.3
Pyrobaculum aerophilum	1.2
Pyrococcus abyssi	1.0

(*Continued*)

Known as:	National Bacteria Average
Pyrococcus furiosus	0.9
Pyrococcus horikoshii	1.2
Ralstonia eutropha	2.1
Ralstonia eutropha	3.6
Ralstonia metallidurans	2.9
Ralstonia solanacearum	2.5
Rhizobium etli	3.0
Rhodobacter sphaeroides	2.0
Rhodococcus sp.	4.4
Rhodoferax ferrireducens	2.4
Rhodopirellula baltica	2.7
Rhodopseudomonas palustris	3.2
Rhodopseudomonas palustris	3.2
Rhodopseudomonas palustris	3.1
Rhodopseudomonas palustris	3.6
Rhodospirillum rubrum	1.9
Rickettsia bellii	0.8
Rickettsia conorii	0.4
Rickettsia felis	0.7
Rickettsia prowazekii	0.3
Rickettsia typhi	0.6
Roseobacter denitrificans	1.9
Rubrobacter xylanophilus	1.6
Saccharophagus degradans	1.6
Salinibacter ruber	1.6
Salmonella enterica	1.3
Salmonella enterica	1.6
Salmonella enterica	1.4
Salmonella enterica	1.4
Salmonella typhimurium	1.9
Shewanella denitrificans	1.2
Shewanella frigidimarina	1.8
Shewanella oneidensis	2.1
Shewanella sp.	1.8
Shewanella sp.	1.8
Shigella boydii	1.4
Shigella dysenteriae	1.3
Shigella flexneri 2a	1.3
Shigella flexneri 2a	1.3
Shigella flexneri 5 str.	1.2
Shigella sonnei	1.5
Silicibacter pomeroyi	1.6

(*Continued*)

Known as:	National Bacteria Average
Silicibacter sp.	1.4
Sinorhizobium meliloti	3.2
Sodalis glossinidius str.	1.2
Sphingopyxis alaskensis	1.6
Staphylococcus aureus	1.4
Staphylococcus aureus	1.4
Staphylococcus aureus	1.4
Staphylococcus aureus	1.4
Staphylococcus aureus	1.2
Staphylococcus aureus	1.4
Staphylococcus aureus	1.3
Staphylococcus aureus	1.2
Staphylococcus aureus	1.2
Staphylococcus epidermidis	1.4
Staphylococcus epidermidis	1.2
Staphylococcus haemolyticus	1.1
Staphylococcus saprophyticus	1.0
Staphylococcus thermophilus	1.1
Streptococcus agalactiae	0.9
Streptococcus agalactiae	0.7
Streptococcus agalactiae	0.8
Streptococcus mutans	0.8
Streptococcus pneumoniae	0.8
Streptococcus pneumoniae	0.9
Streptococcus pyogenes	0.8
Streptococcus pyogenes	0.9
Streptococcus pyogenes	0.9
Streptococcus pyogenes	0.8
Streptococcus pyogenes	0.7
Streptococcus pyogenes	0.8
Streptococcus pyogenes	0.8
Streptococcus pyogenes	0.7
Streptococcus pyogenes	0.8
Streptococcus pyogenes	0.7
Streptococcus thermophilus	1.0
Streptomyces avermitilis	3.6
Streptomyces coelicolor	4.6
Sulfolobus acidocaldarius	0.7
Sulfolobus solfataricus	1.0
Sulfolobus tokodaii	1.2
Symbiobacterium thermophilum	1.6
Synechococcus	1.1

(*Continued*)

Known as:	National Bacteria Average
Synechococcus elongatus	0.8
Synechococcus elongatus	0.8
Synechococcus sp.	1.1
Synechococcus sp.	1.1
Synechococcus sp.	0.8
Synechococcus sp.	1.4
Synechococcus sp.	1.1
Synechocystis	1.4
Syntrophomonas wolfei	0.9
Syntrophus aciditrophicus	1.9
Thermoanaerobacter tengcongensis	1.6
Thermobifida fusca	1.5
Thermococcus kodakarensis	1.1
Thermoplasma acidophilum	0.8
Thermoplasma volcanium	0.8
Thermosynechococcus elongatus	1.3
Thermotoga maritima	1.1
Thermus thermophilus	1.7
Thermus thermophilus	1.8
Thiobacillus denitrificans	1.4
Thiomicrospira crunogena	1.0
Thiomicrospira denitrificans	1.2
Treponema denticola	1.0
Treponema pallidum	0.7
Trichodesmium erythraeum	2.2
Tropheryma whipplei	0.4
Tropheryma whipplei	0.4
Ureaplasma urealyticum	0.4
Vibrio cholerae O1 biovar eltor	1.5
Vibrio fischeri	1.3
Vibrio parahaemolyticus	1.8
Vibrio vulnificus	2.0
Vibrio vulnificus	1.8
Wigglesworthia glossinidia endosymbiont of Glossina brevipalpis	0.5
Wolbachia endosymbiont	0.3
Wolbachia endosymbiont of Drosophila melanogaster	0.6
Wolinella succinogenes	0.8
Xanthomonas axonopodis pv. citri	2.0
Xanthomonas campestris pv. campestris	1.8
Xanthomonas campestris pv. campestris str.	1.9
Xanthomonas campestris pv. vesicatoria str.	1.9

(*Continued*)

Known as:	National Bacteria Average
Xanthomonas oryzae	1.8
Xanthomonas oryzae pv. oryzae	1.6
Xylella fastidiosa	1.3
Xylella fastidiosa	0.9
Yersenia pseudotuberculosis	1.6
Yersinia pestis	1.7
Yersinia pestis	1.6
Yersinia pestis	1.8
Yersinia pestis	1.7
Yersinia pestis biovar Mediaevails	1.7
Zymomonas mobilis	0.8

Appendix B – California Region Average for Bacteria

Listing of bacteria identified and average unique peptides.

Known as:	California Bacteria Average
Acidobacteria bacterium	1.1
Acinetobacter	0.5
Aeropyrum pernix	0.7
Agrobacterium tumefaciens	2.3
Alcanivorax borkumensis	0.9
Alkalilimnicola ehrlichei	1.1
Anabaena variabilis	1.9
Anaeromyxobacter dehalogenans	2.5
Anaplasma marginale	0.4
Anaplasma phagocytophilum	0.4
Aquifex aeolicus	0.8
Archaeoglobus fulgidus	0.7
Azoarcus	1.9
Bacillus anthracis	1.6
Bacillus anthracis	1.7
Bacillus anthracis	1.9
Bacillus cereus	1.8
Bacillus cereus	2.1
Bacillus cereus	2.1
Bacillus clausii	1.9
Bacillus halodurans	1.4
Bacillus licheniformis	1.4
Bacillus subtilis	1.1
Bacillus thuringensis serovar konkukian	2.1
Bacillus thuringiensis	1.9
Bacteroides fragilis	1.3
Bacteroides fragilis	1.3

(*Continued*)

Known as:	California Bacteria Average
Bacteroides thetaiotaomicron	2.6
Bartonella henselae	0.9
Bartonella quintana	0.6
Baumannia cicadellinicola	0.5
Bdellovibrio bacteriovorus	1.1
Bifidobacterium longum	0.4
Bordetella bronchiseptica	2.6
Bordetella parapertussis	2.3
Bordetella pertussis	1.4
Borrelia afzelii	0.8
Borrelia burgdorferi	0.6
Borrelia garinii	1.0
Bradyrhizobium japonicum	4.3
Brucella abortus biovar 1	1.4
Brucella melitensis	1.8
Brucella melitensis biovar Abortus	1.4
Brucella suis	1.4
Buchnera aphidicola	0.4
Burkholderia cenocepacia	3.0
Burkholderia mallei	2.4
Burkholderia pseudomallei	4.3
Burkholderia pseudomallei	2.6
Burkholderia sp.	2.7
Burkholderia thailandensis	2.5
Burkholderia xenovorans	2.4
Campylobacter jejuni	0.7
Campylobacter jejuni	0.9
Candidatus Blochmannia floridanus	0.2
Candidatus Blochmannia pennsylvanicus str.	0.1
Candidatus Pelagibacter ubique	0.6
Candidatus Protochlamydia amoebophila	0.5
Carboxydothermus hydrogenoformans	0.9
Caulobacter crescentus	2.1
Chlamydia muridarum	0.4
Chlamydia trachomatis	0.4
Chlamydia trachomatis	0.4
Chlamydophila abortus	0.6
Chlamydophila caviae	0.8
Chlamydophila felis	0.9
Chlamydophila pneumoniae	0.4
Chlamydophila pneumoniae	0.4
Chlamydophila pneumoniae	0.4

(Continued)

Known as:	California Bacteria Average
Chlamydophila pneumoniae	0.5
Chlorobium chlorochromatii	0.6
Chlorobium tepidum	0.6
Chromobacterium violaceum	1.8
Chromohalobacter salexigens	1.4
Clostridium acetobutylicum	1.0
Clostridium perfringens	0.9
Clostridium perfringens	0.8
Clostridium perfringens	0.6
Clostridium tetani	1.6
Colwellia psychrerythraea	1.4
Corynebacterium diphtheriae	0.6
Corynebacterium efficiens	0.8
Corynebacterium glutamicum	0.9
Corynebacterium jeikeium	0.9
Coxiella burnetii	0.4
Cytophaga hutchinsonii	1.2
Dechloromonas aromatica	1.4
Dehalococcoides ethenogenes	1.7
Dehalococcoides sp.	0.3
Deinococcus geothermalis	1.0
Deinococcus radiodurans	1.6
Desulfitobacterium hafniense	1.5
Desulfotalea psychrophila	1.1
Desulfovibrio desulfuricans	1.6
Desulfovibrio vulgaris	1.5
Ehrlichia canis str.	0.6
Ehrlichia chaffeensis str.	0.6
Ehrlichia ruminantium	0.4
Ehrlichia ruminantium	0.4
Enterococcus faecalis	1.1
Erwinia carotovoa	0.9
Erythrobacter litoralis	1.0
Escherichia coli	2.0
Escherichia coli	1.9
Escherichia coli	2.0
Escherichia coli	2.0
Escherichia coli	1.9
Escherichia coli	1.9
Escherichia coli	2.0
Escherichia coli	2.0
Fowlpox Virus	0.2

(*Continued*)

Known as:	California Bacteria Average
Francisella tularensis	0.9
Francisella tularensis	0.9
Francisella tularensis	0.9
Francisella tularensis	0.9
Frankia alni	3.6
Frankia sp.	2.2
Fusobacterium nucleatum	0.7
Geobacillus kausophilus	1.3
Geobacter metallireducens	1.1
Geobacter sulfurreducens	1.4
Gloeobacter violaceus	1.9
Gluconobacter oxydans	1.4
Granulibacter bethesdensis	0.8
Haemophilus ducreyi	0.9
Haemophilus influenzae	0.8
Haemophilus influenzae	0.6
Haemophilus somnus	0.7
Hahella chejuensis	2.4
Haloarcula marismortui	1.0
Halobacterium	0.8
Haloquadratum walsbyi	1.1
Helicobacter acinonychis str.	0.4
Helicobacter hepaticus	0.4
Helicobacter pylori	1.1
Helicobacter pylori	0.7
Helicobacter pylori	0.6
Hyphomonas neptunium	1.5
Idiomarina loihiensis	0.9
Jannaschia sp.	1.8
Klebsiella pneumoniae	2.0
Lactobacillus acidophilus	1.0
Lactobacillus delbrueckii	0.6
Lactobacillus johnsonii	0.8
Lactobacillus plantarum	0.9
Lactobacillus sakei	1.0
Lactobacillus salivarius	0.4
Lactococcus lactis	0.9
Lawsonia intracellularis	0.6
Legionella pneumophila	1.3
Legionella pneumophila	1.3
Legionella pneumophila	1.3
Leifsonia xyli	0.6

(Continued)

Known as:	California Bacteria Average
Leptospira interrogans serovar Copenhageni	1.5
Leptospira interrogans serovar lai	1.7
Listeria innocua	1.0
Listeria monocytogenes	1.2
Listeria monocytogenes	0.7
Magnetospirillum magneticum	2.2
Mannheimia succiniproducens	0.7
Maricaulis maris	0.9
Mesoplasma florum	0.3
Mesorhizobium loti	3.4
Mesorhizobium sp.	1.9
Methanocaldococcus jannaschii	1.3
Methanococcoides burtonii	1.4
Methanococcus maripaludis	0.6
Methanopyrus kandleri	0.6
Methanosarcina acetivorans	1.9
Methanosarcina barkeri str.	1.3
Methanosarcina mazei	2.0
Methanosphaera stadtmanae	0.4
Methanospirillum hungatei	1.1
Methanothermobacter thermautotrophicus	0.4
Methylobacillus flagellatus	0.9
Methylococcus capsulatus	1.5
Moorella thermoacetica	1.2
Mycobacterium avium	1.6
Mycobacterium bovis	1.5
Mycobacterium leprae	0.8
Mycobacterium sp.	2.0
Mycobacterium tuberculosis	1.5
Mycobacterium tuberculosis	1.5
Mycoplasma gallisepticum	0.4
Mycoplasma genitalium	0.1
Mycoplasma hyopneumoniae	0.3
Mycoplasma hyopneumoniae	0.4
Mycoplasma hyopneumoniae	0.3
Mycoplasma mobile	0.3
Mycoplasma mycoides	0.1
Mycoplasma penetrans	0.3
Mycoplasma pneumoniae	0.1
Mycoplasma pulmonis	0.3
Mycoplasma synoviae	0.5
Myxococcus xanthus	3.1

(*Continued*)

Known as:	California Bacteria Average
Natronomonas pharaonis	0.9
Neisseria gonorrhoeae	0.6
Neisseria meningitidis	0.9
Neisseria meningitidis	0.6
Neorickettsia sennetsu str.	0.3
Nitrobacter hamburgensis	2.4
Nitrobacter winogradskyi	1.0
Nitrosococcus oceani	0.7
Nitrosomonas europaea	0.6
Nitrosomonas eutropha	1.4
Nitrosospira multiformis	1.1
Nocardia farcinica	2.4
Nostoc sp.	1.8
Novosphingobium aromaticivorans	1.5
Oceanobacillus iheyensis	0.6
Onion Yellows Phytoplasma	0.1
Pasteurella Multocida	0.9
Pelobacter carbinolicus	1.3
Pelodictyon luteolum	1.1
Photobacterium profundum	1.8
Photorhabdus luminescens	1.9
Picrophilus torridus	0.8
Polaromonas sp.	1.8
Porphyromonas gingivalis	0.9
Prochlorococcus marinus	0.6
Prochlorococcus marinus	0.9
Prochlorococcus marinus	1.4
Prochlorococcus marinus str.	1.1
Prochlorococcus marinus str.	0.9
Propionibacterium acnes	0.9
Pseudoalteromonas atlantica	2.1
Pseudoalteromonas haloplanktis	1.3
Pseudomonas aeruginosa	2.4
Pseudomonas entomophila	1.7
Pseudomonas fluorescens	2.4
Pseudomonas fluorescens	2.0
Pseudomonas putida	1.6
Pseudomonas syringae pv. phaseolicola	1.0
Pseudomonas syringae pv. syringae	1.4
Pseudomonas syringae pv. tomato	1.1
Psychrobacter arcticus	0.7
Psychrobacter cryohalolentis	0.8

(*Continued*)

Known as:	California Bacteria Average
Pyrobaculum aerophilum	0.9
Pyrococcus abyssi	0.7
Pyrococcus furiosus	1.2
Pyrococcus horikoshii	1.1
Ralstonia eutropha	1.4
Ralstonia eutropha	2.4
Ralstonia metallidurans	2.7
Ralstonia solanacearum	1.9
Rhizobium etli	2.9
Rhodobacter sphaeroides	1.3
Rhodococcus sp.	3.1
Rhodoferax ferrireducens	1.9
Rhodopirellula baltica	2.5
Rhodopseudomonas palustris	2.8
Rhodopseudomonas palustris	2.4
Rhodopseudomonas palustris	2.3
Rhodopseudomonas palustris	2.5
Rhodospirillum rubrum	2.1
Rickettsia bellii	0.6
Rickettsia conorii	0.6
Rickettsia felis	0.5
Rickettsia prowazekii	0.5
Rickettsia typhi	0.4
Roseobacter denitrificans	1.5
Rubrobacter xylanophilus	1.4
Saccharophagus degradans	1.6
Salinibacter ruber	0.8
Salmonella enterica	1.7
Salmonella enterica	1.9
Salmonella enterica	1.6
Salmonella enterica	1.9
Salmonella typhimurium	2.1
Shewanella denitrificans	2.0
Shewanella frigidimarina	0.9
Shewanella oneidensis	1.0
Shewanella sp.	1.6
Shewanella sp.	1.7
Shigella boydii	1.4
Shigella dysenteriae	1.0
Shigella flexneri 2a	1.4
Shigella flexneri 2a	1.4
Shigella flexneri 5 str.	1.4

(*Continued*)

Known as:	California Bacteria Average
Shigella sonnei	1.7
Silicibacter pomeroyi	1.8
Silicibacter sp.	1.2
Sinorhizobium meliloti	2.2
Sodalis glossinidius str.	0.7
Sphingopyxis alaskensis	2.1
Staphylococcus aureus	1.1
Staphylococcus aureus	1.4
Staphylococcus aureus	0.9
Staphylococcus aureus	0.9
Staphylococcus aureus	0.9
Staphylococcus aureus	1.0
Staphylococcus aureus	0.9
Staphylococcus aureus	1.0
Staphylococcus aureus	1.0
Staphylococcus epidermidis	1.0
Staphylococcus epidermidis	1.0
Staphylococcus haemolyticus	1.2
Staphylococcus saprophyticus	1.1
Staphylococcus thermophilus	0.5
Streptococcus agalactiae	0.9
Streptococcus agalactiae	0.8
Streptococcus agalactiae	0.7
Streptococcus mutans	0.4
Streptococcus pneumoniae	0.4
Streptococcus pneumoniae	0.6
Streptococcus pyogenes	0.6
Streptococcus pyogenes	0.5
Streptococcus pyogenes	0.6
Streptococcus pyogenes	0.6
Streptococcus pyogenes	0.5
Streptococcus pyogenes	0.6
Streptococcus pyogenes	0.6
Streptococcus pyogenes	0.4
Streptococcus pyogenes	0.5
Streptococcus pyogenes	0.4
Streptococcus pyogenes	0.6
Streptococcus thermophilus	0.4
Streptomyces avermitilis	4.3
Streptomyces coelicolor	3.4
Sulfolobus acidocaldarius	0.6
Sulfolobus solfataricus	0.7

(*Continued*)

Known as:	California Bacteria Average
Sulfolobus tokodaii	1.1
Symbiobacterium thermophilum	1.5
Synechococcus	0.7
Synechococcus elongatus	0.8
Synechococcus elongatus	0.9
Synechococcus sp.	0.9
Synechococcus sp.	1.1
Synechococcus sp.	0.6
Synechococcus sp.	0.6
Synechococcus sp.	0.9
Synechocystis	1.1
Syntrophomonas wolfei	1.2
Syntrophus aciditrophicus	1.0
Thermoanaerobacter tengcongensis	1.4
Thermobifida fusca	0.9
Thermococcus kodakarensis	0.6
Thermoplasma acidophilum	0.7
Thermoplasma volcanium	0.6
Thermosynechococcus elongatus	0.4
Thermotoga maritima	0.9
Thermus thermophilus	1.1
Thermus thermophilus	0.6
Thiobacillus denitrificans	1.4
Thiomicrospira crunogena	0.6
Thiomicrospira denitrificans	1.0
Treponema denticola	1.1
Treponema pallidum	0.4
Trichodesmium erythraeum	2.1
Tropheryma whipplei	0.5
Tropheryma whipplei	0.6
Ureaplasma urealyticum	0.2
Vibrio cholerae O1 biovar eltor	1.5
Vibrio fischeri	1.6
Vibrio parahaemolyticus	1.7
Vibrio vulnificus	1.9
Vibrio vulnificus	1.8
Wigglesworthia glossinidia endosymbiont of Glossina brevipalpis	0.4
Wolbachia endosymbiont	0.1
Wolbachia endosymbiont of Drosophila melanogaster	0.7
Wolinella succinogenes	0.8

(*Continued*)

Known as:	California Bacteria Average
Xanthomonas axonopodis pv. citri	1.4
Xanthomonas campestris pv. campestris	1.7
Xanthomonas campestris pv. campestris str.	1.9
Xanthomonas campestris pv. vesicatoria str.	1.6
Xanthomonas oryzae	1.1
Xanthomonas oryzae pv. oryzae	1.3
Xylella fastidiosa	0.9
Xylella fastidiosa	1.1
Yersenia pseudotuberculosis	1.3
Yersinia pestis	1.1
Yersinia pestis	1.1
Yersinia pestis	1.1
Yersinia pestis	1.1
Yersinia pestis biovar Mediaevails	1.1
Zymomonas mobilis	0.6

Appendix C – Florida Region Average for Bacteria

Listing of bacteria detected and identified along with average number of unique peptides.

Known as:	Florida bacteria average
Acidobacteria bacterium	1.4
Acinetobacter	0.6
Aeropyrum pernix	0.5
Agrobacterium tumefaciens	1.0
Alcanivorax borkumensis	0.7
Alkalilimnicola ehrlichei	0.6
Anabaena variabilis	0.9
Anaeromyxobacter dehalogenans	1.5
Anaplasma marginale	0.2
Anaplasma phagocytophilum	0.1
Aquifex aeolicus	0.4
Archaeoglobus fulgidus	0.4
Azoarcus	0.8
Bacillus anthracis	0.9
Bacillus anthracis	0.9
Bacillus anthracis	1.0
Bacillus cereus	1.0
Bacillus cereus	1.1
Bacillus cereus	1.1
Bacillus clausii	0.8
Bacillus halodurans	0.4
Bacillus licheniformis	1.0
Bacillus subtilis	0.9
Bacillus thuringensis serovar konkukian	1.0
Bacillus thuringiensis	1.0
Bacteroides fragilis	0.9

(*Continued*)

Known as:	Florida bacteria average
Bacteroides fragilis	1.1
Bacteroides thetaiotaomicron	1.5
Bartonella henselae	0.4
Bartonella quintana	0.5
Baumannia cicadellinicola	0.1
Bdellovibrio bacteriovorus	0.5
Bifidobacterium longum	0.2
Bordetella bronchiseptica	0.9
Bordetella parapertussis	0.8
Bordetella pertussis	0.9
Borrelia afzelii	0.4
Borrelia burgdorferi	0.4
Borrelia garinii	0.3
Bradyrhizobium japonicum	1.9
Brucella abortus biovar 1	1.0
Brucella melitensis	1.0
Brucella melitensis biovar Abortus	0.9
Brucella suis	0.9
Buchnera aphidicola	0.1
Buchnera aphidicola	0.2
Burkholderia cenocepacia	1.4
Burkholderia mallei	0.9
Burkholderia pseudomallei	1.6
Burkholderia pseudomallei	1.0
Burkholderia sp.	1.5
Burkholderia thailandensis	1.1
Burkholderia xenovorans	1.8
Campylobacter jejuni	0.1
Candidatus Blochmannia floridanus	0.2
Candidatus Blochmannia pennsylvanicus str.	0.2
Candidatus Pelagibacter ubique	0.6
Candidatus Protochlamydia amoebophila	0.3
Carboxydothermus hydrogenoformans	0.4
Caulobacter crescentus	1.5
Chlamydia muridarum	0.1
Chlamydia trachomatis	0.1
Chlamydia trachomatis	0.1
Chlamydophila abortus	0.1
Chlamydophila caviae	0.2
Chlamydophila felis	0.2
Chlamydophila pneumoniae	0.4

(*Continued*)

Known as:	Florida bacteria average
Chlamydophila pneumoniae	0.3
Chlamydophila pneumoniae	0.3
Chlamydophila pneumoniae	0.3
Chlorobium chlorochromatii	0.1
Chlorobium tepidum	0.5
Chromobacterium violaceum	1.0
Chromohalobacter salexigens	0.6
Clostridium acetobutylicum	0.7
Clostridium perfringens	0.6
Clostridium perfringens	0.6
Clostridium perfringens	0.5
Clostridium tetani	0.5
Colwellia psychrerythraea	0.8
Corynebacterium diphtheriae	0.6
Corynebacterium efficiens	0.5
Corynebacterium glutamicum	0.5
Corynebacterium jeikeium	0.6
Coxiella burnetii	0.3
Cytophaga hutchinsonii	0.8
Dechloromonas aromatica	1.0
Dehalococcoides ethenogenes	1.1
Dehalococcoides sp.	0.3
Deinococcus geothermalis	0.6
Deinococcus radiodurans	0.5
Desulfitobacterium hafniense	0.9
Desulfotalea psychrophila	0.6
Desulfovibrio desulfuricans	0.6
Desulfovibrio vulgaris	0.8
Ehrlichia canis str.	0.1
Ehrlichia chaffeensis str.	0.1
Enterococcus faecalis	0.7
Erwinia carotovoa	0.9
Erythrobacter litoralis	0.9
Escherichia coli	0.8
Escherichia coli	0.8
Escherichia coli	1.0
Escherichia coli	0.8
Escherichia coli	1.1
Escherichia coli	1.0
Escherichia coli	1.0
Escherichia coli	0.7
Fowlpox Virus	0.1

(*Continued*)

Known as:	Florida bacteria average
Francisella tularensis	0.4
Francisella tularensis	0.4
Francisella tularensis	0.4
Francisella tularensis	0.4
Frankia alni	1.4
Frankia sp.	1.0
Fusobacterium nucleatum	0.4
Geobacillus kausophilus	0.8
Geobacter metallireducens	1.0
Geobacter sulfurreducens	0.8
Gloeobacter violaceus	1.2
Gluconobacter oxydans	0.3
Granulibacter bethesdensis	0.3
Haemophilus ducreyi	0.2
Haemophilus influenzae	0.4
Haemophilus influenzae	0.3
Haemophilus somnus	0.2
Hahella chejuensis	1.1
Haloarcula marismortui	0.6
Halobacterium	0.4
Haloquadratum walsbyi	0.3
Helicobacter acinonychis str.	0.2
Helicobacter hepaticus	0.4
Helicobacter pylori	0.5
Helicobacter pylori	0.1
Helicobacter pylori	0.5
Hyphomonas neptunium	0.9
Idiomarina loihiensis	0.5
Jannaschia sp.	1.1
Klebsiella pneumoniae	1.3
Lactobacillus acidophilus	0.6
Lactobacillus delbrueckii	0.2
Lactobacillus johnsonii	0.3
Lactobacillus plantarum	0.5
Lactobacillus sakei	0.2
Lactobacillus salivarius	0.3
Lactococcus lactis	0.5
Lawsonia intracellularis	0.2
Legionella pneumophila	0.9
Legionella pneumophila	1.0
Legionella pneumophila	1.0
Leifsonia xyli	0.3

(*Continued*)

Known as:	Florida bacteria average
Leptospira interrogans serovar Copenhageni	0.6
Leptospira interrogans serovar lai	0.8
Listeria innocua	0.4
Listeria monocytogenes	0.3
Listeria monocytogenes	0.5
Magnetospirillum magneticum	1.0
Mannheimia succiniproducens	0.5
Maricaulis maris	0.6
Mesoplasma florum	0.2
Mesorhizobium loti	1.5
Mesorhizobium sp.	0.8
Methanocaldococcus jannaschii	0.3
Methanococcoides burtonii	0.4
Methanococcus maripaludis	0.1
Methanopyrus kandleri	0.2
Methanosarcina acetivorans	0.6
Methanosarcina barkeri str.	0.6
Methanosarcina mazei	0.7
Methanosphaera stadtmanae	0.3
Methanospirillum hungatei	0.7
Methanothermobacter thermautotrophicus	0.6
Methylobacillus flagellatus	0.7
Methylococcus capsulatus	0.5
Moorella thermoacetica	0.7
Mycobacterium avium	0.8
Mycobacterium bovis	0.7
Mycobacterium leprae	0.5
Mycobacterium sp.	1.1
Mycobacterium tuberculosis	0.9
Mycobacterium tuberculosis	0.7
Mycoplasma capricolum	0.4
Mycoplasma gallisepticum	0.4
Mycoplasma genitalium	0.2
Mycoplasma hyopneumoniae	0.2
Mycoplasma hyopneumoniae	0.1
Mycoplasma hyopneumoniae	0.2
Mycoplasma mobile	0.1
Mycoplasma mycoides	0.3
Mycoplasma penetrans	0.4
Mycoplasma pneumoniae	0.1

(*Continued*)

Known as:	Florida bacteria average
Mycoplasma pulmonis	0.2
Mycoplasma synoviae	0.1
Myxococcus xanthus	2.3
Nanoarchaeum equitans	0.1
Natronomonas pharaonis	0.8
Neisseria gonorrhoeae	0.4
Neisseria meningitidis	0.4
Neisseria meningitidis	0.4
Neorickettsia sennetsu str.	0.1
Nitrobacter hamburgensis	1.1
Nitrobacter winogradskyi	0.9
Nitrosococcus oceani	0.6
Nitrosomonas europaea	0.6
Nitrosomonas eutropha	0.4
Nitrosospira multiformis	0.3
Nocardia farcinica	1.1
Nostoc sp.	1.0
Novosphingobium aromaticivorans	1.1
Oceanobacillus iheyensis	0.3
Onion Yellows Phytoplasma	0.2
Pasteurella Multocida	0.3
Pelobacter carbinolicus	0.7
Pelodictyon luteolum	0.6
Photobacterium profundum	1.0
Photorhabdus luminescens	0.7
Picrophilus torridus	0.1
Polaromonas sp.	1.1
Porphyromonas gingivalis	0.3
Prochlorococcus marinus	0.5
Prochlorococcus marinus	0.3
Prochlorococcus marinus	0.4
Prochlorococcus marinus str.	0.3
Prochlorococcus marinus str.	0.2
Propionibacterium acnes	0.7
Pseudoalteromonas atlantica	0.9
Pseudoalteromonas haloplanktis	0.6
Pseudomonas aeruginosa	1.1
Pseudomonas entomophila	1.0
Pseudomonas fluorescens	1.2
Pseudomonas fluorescens	1.2
Pseudomonas putida	1.0
Pseudomonas syringae pv. phaseolicola	1.1

(*Continued*)

Known as:	Florida bacteria average
Pseudomonas syringae pv. syringae	0.9
Pseudomonas syringae pv. tomato	1.2
Psychrobacter arcticus	0.5
Psychrobacter cryohalolentis	0.4
Pyrobaculum aerophilum	0.5
Pyrococcus abyssi	0.3
Pyrococcus furiosus	0.3
Pyrococcus horikoshii	0.5
Ralstonia eutropha	0.8
Ralstonia eutropha	1.2
Ralstonia metallidurans	1.5
Ralstonia solanacearum	1.3
Rhizobium etli	1.0
Rhodobacter sphaeroides	1.2
Rhodococcus sp.	1.9
Rhodoferax ferrireducens	1.0
Rhodopirellula baltica	1.9
Rhodopseudomonas palustris	1.0
Rhodopseudomonas palustris	1.0
Rhodopseudomonas palustris	1.6
Rhodopseudomonas palustris	1.5
Rhodospirillum rubrum	0.9
Rickettsia bellii	0.1
Rickettsia conorii	0.2
Rickettsia felis	0.3
Rickettsia prowazekii	0.1
Rickettsia typhi	0.2
Roseobacter denitrificans	0.9
Rubrobacter xylanophilus	0.8
Saccharophagus degradans	0.7
Salinibacter ruber	0.7
Salmonella enterica	0.8
Salmonella enterica	0.8
Salmonella enterica	0.8
Salmonella enterica	0.9
Salmonella typhimurium	1.0
Shewanella denitrificans	0.8
Shewanella frigidimarina	1.0
Shewanella oneidensis	0.7
Shewanella sp.	0.7
Shewanella sp.	0.7
Shigella boydii	0.6

(*Continued*)

Known as:	Florida bacteria average
Shigella dysenteriae	0.7
Shigella flexneri 2a	0.6
Shigella flexneri 2a	0.8
Shigella flexneri 5 str.	0.7
Shigella sonnei	0.7
Silicibacter pomeroyi	0.7
Silicibacter sp.	1.1
Sinorhizobium meliloti	1.6
Sodalis glossinidius str.	0.3
Sphingopyxis alaskensis	0.8
Staphylococcus aureus	0.6
Staphylococcus aureus	0.6
Staphylococcus aureus	0.7
Staphylococcus aureus	0.6
Staphylococcus aureus	0.6
Staphylococcus aureus	0.6
Staphylococcus aureus	0.6
Staphylococcus aureus	0.7
Staphylococcus aureus	0.5
Staphylococcus epidermidis	0.7
Staphylococcus epidermidis	0.4
Staphylococcus haemolyticus	0.4
Staphylococcus saprophyticus	0.3
Staphylococcus thermophilus	0.6
Streptococcus agalactiae	0.3
Streptococcus agalactiae	0.2
Streptococcus agalactiae	0.2
Streptococcus mutans	0.4
Streptococcus pneumoniae	0.3
Streptococcus pneumoniae	0.5
Streptococcus pyogenes	0.6
Streptococcus pyogenes	0.7
Streptococcus pyogenes	0.6
Streptococcus pyogenes	0.7
Streptococcus pyogenes	0.7
Streptococcus pyogenes	0.6
Streptococcus pyogenes	0.7
Streptococcus pyogenes	0.6
Streptococcus pyogenes	0.7
Streptococcus pyogenes	0.6
Streptococcus pyogenes	0.6
Streptococcus thermophilus	0.6

(*Continued*)

Known as:	Florida bacteria average
Streptomyces avermitilis	1.9
Streptomyces coelicolor	2.4
Sulfolobus acidocaldarius	0.5
Sulfolobus solfataricus	0.5
Sulfolobus tokodaii	0.6
Symbiobacterium thermophilum	0.7
Synechococcus	0.5
Synechococcus elongatus	0.3
Synechococcus elongatus	0.4
Synechococcus sp.	0.7
Synechococcus sp.	0.7
Synechococcus sp.	0.6
Synechococcus sp.	0.4
Synechococcus sp.	0.7
Synechocystis	1.0
Syntrophomonas wolfei	0.4
Syntrophus aciditrophicus	0.8
Thermoanaerobacter tengcongensis	0.7
Thermobifida fusca	0.8
Thermococcus kodakarensis	0.7
Thermoplasma acidophilum	0.2
Thermoplasma volcanium	0.2
Thermosynechococcus elongatus	0.3
Thermotoga maritima	0.4
Thermus thermophilus	0.3
Thermus thermophilus	0.4
Thiobacillus denitrificans	0.8
Thiomicrospira crunogena	0.4
Thiomicrospira denitrificans	0.6
Treponema denticola	0.7
Treponema pallidum	0.3
Trichodesmium erythraeum	0.7
Tropheryma whipplei	0.1
Vibrio cholerae O1 biovar eltor	0.7
Vibrio fischeri	1.0
Vibrio parahaemolyticus	1.0
Vibrio vulnificus	0.7
Vibrio vulnificus	0.8
Wigglesworthia glossinidia endosymbiont of *Glossina brevipalpis*	0.1
Wolbachia endosymbiont	0.1

(*Continued*)

Known as:	**Florida bacteria average**
Wolbachia endosymbiont *Drosophila melanogaster*	0.2
Wolinella succinogenes	0.4
Xanthomonas axonopodis pv. citri	1.1
Xanthomonas campestris pv. campestris	1.1
Xanthomonas campestris pv. campestris str.	1.2
Xanthomonas campestris pv. vesicatoria str.	1.1
Xanthomonas oryzae	0.7
Xanthomonas oryzae pv. oryzae	0.5
Xylella fastidiosa	0.7
Xylella fastidiosa	0.4
Yersenia pseudotuberculosis	0.9
Yersinia pestis	0.7
Yersinia pestis	0.7
Yersinia pestis	0.7
Yersinia pestis	0.7
Yersinia pestis biovar Mediaevails	0.7
Zymomonas mobilis	0.5

Appendix D – Idaho Region Average for Bacteria

Listing of Bacteria detected and identified along with average number of unique peptides.

Known as:	Idaho Bacteria Average
Acidobacteria bacterium	3.3
Acinetobacter	0.5
Aeropyrum pernix	1.3
Agrobacterium tumefaciens	1.8
Alcanivorax borkumensis	1.3
Alkalilimnicola ehrlichei	1.0
Anabaena variabilis	2.8
Anaeromyxobacter dehalogenans	2.3
Anaplasma marginale	1.0
Aquifex aeolicus	0.8
Archaeoglobus fulgidus	1.3
Azoarcus	2.0
Bacillus anthracis	1.5
Bacillus anthracis	1.5
Bacillus anthracis	1.8
Bacillus cereus	2.3
Bacillus cereus	2.5
Bacillus cereus	1.8
Bacillus clausii	1.8
Bacillus halodurans	0.8
Bacillus licheniformis	1.5
Bacillus subtilis	2.0
Bacillus thuringensis serovar konkukian	3.0
Bacillus thuringiensis	2.0
Bacteroides fragilis	1.8
Bacteroides fragilis	2.0
Bacteroides thetaiotaomicron	2.0
Bartonella henselae	0.5

(*Continued*)

Known as:	Idaho Bacteria Average
Bartonella quintana	1.0
Bdellovibrio bacteriovorus	1.0
Bifidobacterium longum	0.5
Bordetella bronchiseptica	1.5
Bordetella parapertussis	1.0
Bordetella pertussis	0.8
Borrelia afzelii	0.8
Borrelia burgdorferi	1.0
Borrelia garinii	1.0
Bradyrhizobium japonicum	2.8
Brucella abortus biovar 1	1.0
Brucella melitensis	1.0
Brucella melitensis biovar Abortus	0.8
Brucella suis	0.8
Burkholderia cenocepacia	1.8
Burkholderia mallei	2.0
Burkholderia pseudomallei	4.3
Burkholderia pseudomallei	2.0
Burkholderia sp.	2.8
Burkholderia thailandensis	2.3
Burkholderia xenovorans	2.3
Campylobacter jejuni	0.5
Candidatus Pelagibacter ubique	0.8
Candidatus Protochlamydia amoebophila	1.3
Carboxydothermus hydrogenoformans	1.5
Caulobacter crescentus	1.8
Chlamydia muridarum	1.3
Chlamydia trachomatis	1.0
Chlamydia trachomatis	1.0
Chlamydophila abortus	0.8
Chlamydophila caviae	1.0
Chlamydophila felis	0.8
Chlamydophila pneumoniae	0.8
Chlamydophila pneumoniae	0.8
Chlamydophila pneumoniae	0.8
Chlamydophila pneumoniae	0.8
Chlorobium chlorochromatii	0.8
Chlorobium tepidum	0.8
Chromobacterium violaceum	1.8
Chromohalobacter salexigens	2.8
Clostridium acetobutylicum	1.0
Clostridium perfringens	1.0

(Continued)

Known as:	Idaho Bacteria Average
Clostridium perfringens	0.8
Clostridium perfringens	1.0
Clostridium tetani	1.8
Colwellia psychrerythraea	2.8
Corynebacterium diphtheriae	0.5
Corynebacterium efficiens	0.5
Corynebacterium jeikeium	0.5
Cytophaga hutchinsonii	2.3
Dechloromonas aromatica	1.8
Dehalococcoides ethenogenes	3.3
Dehalococcoides sp.	0.8
Deinococcus geothermalis	1.0
Deinococcus radiodurans	1.0
Desulfitobacterium hafniense	2.5
Desulfotalea psychrophila	1.3
Desulfovibrio desulfuricans	2.3
Desulfovibrio vulgaris	1.0
Ehrlichia ruminantium	0.5
Ehrlichia ruminantium	0.5
Enterococcus faecalis	0.5
Erwinia carotovoa	1.3
Erythrobacter litoralis	1.3
Escherichia coli	1.3
Escherichia coli	1.0
Escherichia coli	1.3
Escherichia coli	1.0
Escherichia coli	1.8
Escherichia coli	1.8
Escherichia coli	1.0
Escherichia coli	1.0
Frankia alni	3.8
Frankia sp.	2.0
Fusobacterium nucleatum	1.0
Geobacillus kausophilus	1.0
Geobacter metallireducens	1.5
Geobacter sulfurreducens	1.3
Gloeobacter violaceus	1.3
Granulibacter bethesdensis	1.0
Haemophilus ducreyi	0.8
Hahella chejuensis	2.0
Halobacterium	0.5
Haloquadratum walsbyi	0.8

(*Continued*)

Known as:	Idaho Bacteria Average
Helicobacter acinonychis str.	0.8
Helicobacter hepaticus	0.5
Helicobacter pylori	0.8
Helicobacter pylori	0.8
Helicobacter pylori	1.0
Idiomarina loihiensis	1.0
Jannaschia sp.	0.8
Klebsiella pneumoniae	2.0
Lactobacillus acidophilus	1.0
Lactobacillus delbrueckii	0.5
Lactobacillus johnsonii	1.0
Lactobacillus plantarum	1.0
Lactobacillus sakei	0.8
Lactobacillus salivarius	0.5
Lactococcus lactis	1.3
Lawsonia intracellularis	0.5
Legionella pneumophila	0.8
Legionella pneumophila	0.8
Legionella pneumophila	1.5
Leifsonia xyli	0.8
Leptospira interrogans serovar Copenhageni	1.0
Leptospira interrogans serovar lai	0.8
Listeria innocua	1.3
Listeria monocytogenes	1.3
Listeria monocytogenes	1.0
Magnetospirillum magneticum	1.3
Mannheimia succiniproducens	0.5
Maricaulis maris	1.5
Mesoplasma florum	0.5
Mesorhizobium loti	3.0
Mesorhizobium sp.	1.5
Methanocaldococcus jannaschii	1.8
Methanococcoides burtonii	1.8
Methanococcus maripaludis	0.5
Methanopyrus kandleri	0.8
Methanosarcina acetivorans	1.0
Methanosarcina barkeri str.	1.3
Methanosarcina mazei	1.0
Methanosphaera stadtmanae	0.8
Methanospirillum hungatei	1.5

(*Continued*)

Known as:	Idaho Bacteria Average
Methanothermobacter thermautotrophicus	0.5
Methylobacillus flagellatus	1.0
Methylococcus capsulatus	0.8
Mycobacterium avium	1.5
Mycobacterium leprae	0.5
Mycobacterium sp.	1.3
Mycobacterium tuberculosis	0.8
Mycoplasma hyopneumoniae	0.5
Mycoplasma penetrans	1.0
Myxococcus xanthus	4.3
Nanoarchaeum equitans	0.5
Neisseria gonorrhoeae	1.5
Neisseria meningitidis	1.8
Neisseria meningitidis	1.5
Nitrobacter hamburgensis	1.8
Nitrobacter winogradskyi	1.8
Nitrosococcus oceani	1.8
Nitrosomonas europaea	1.0
Nitrosomonas eutropha	1.0
Nitrosospira multiformis	1.8
Nocardia farcinica	2.8
Nostoc sp.	2.3
Novosphingobium aromaticivorans	1.5
Oceanobacillus iheyensis	1.3
Onion Yellows Phytoplasma	0.5
Pasteurella Multocida	1.5
Pelobacter carbinolicus	0.5
Pelodictyon luteolum	0.5
Photobacterium profundum	0.8
Photorhabdus luminescens	2.3
Polaromonas sp.	2.8
Porphyromonas gingivalis	0.5
Prochlorococcus marinus	0.5
Prochlorococcus marinus	0.5
Prochlorococcus marinus	1.8
Prochlorococcus marinus str.	0.5
Prochlorococcus marinus str.	0.5
Propionibacterium acnes	1.3
Pseudoalteromonas atlantica	1.0
Pseudoalteromonas haloplanktis	0.8
Pseudomonas aeruginosa	2.8

(*Continued*)

Known as:	Idaho Bacteria Average
Pseudomonas entomophila	1.3
Pseudomonas fluorescens	2.8
Pseudomonas fluorescens	1.3
Pseudomonas putida	1.3
Pseudomonas syringae pv. phaseolicola	2.0
Pseudomonas syringae pv. syringae	1.5
Pseudomonas syringae pv. tomato	2.8
Psychrobacter arcticus	0.5
Pyrobaculum aerophilum	1.5
Pyrococcus abyssi	0.8
Pyrococcus furiosus	1.0
Pyrococcus horikoshii	1.3
Ralstonia eutropha	0.8
Ralstonia eutropha	1.8
Ralstonia metallidurans	1.8
Ralstonia solanacearum	1.3
Rhizobium etli	1.3
Rhodobacter sphaeroides	1.5
Rhodococcus sp.	3.5
Rhodoferax ferrireducens	1.3
Rhodopirellula baltica	1.0
Rhodopseudomonas palustris	1.8
Rhodopseudomonas palustris	1.0
Rhodopseudomonas palustris	1.5
Rhodopseudomonas palustris	2.0
Rhodospirillum rubrum	1.5
Rickettsia bellii	0.8
Rickettsia felis	0.8
Rickettsia prowazekii	0.5
Rickettsia typhi	0.5
Roseobacter denitrificans	0.8
Rubrobacter xylanophilus	1.3
Saccharophagus degradans	1.5
Salinibacter ruber	1.3
Salmonella enterica	1.0
Salmonella enterica	1.8
Salmonella enterica	1.5
Salmonella enterica	1.5
Salmonella typhimurium	2.3
Shewanella denitrificans	3.0
Shewanella frigidimarina	1.5
Shewanella oneidensis	2.3

(Continued)

Known as:	Idaho Bacteria Average
Shewanella sp.	2.3
Shewanella sp.	2.3
Shigella boydii	1.3
Shigella dysenteriae	1.5
Shigella flexneri 2a	1.3
Shigella flexneri 2a	1.3
Shigella flexneri 5 str.	1.5
Shigella sonnei	1.3
Silicibacter pomeroyi	1.0
Silicibacter sp.	1.5
Sinorhizobium meliloti	2.8
Sodalis glossinidius str.	0.8
Sphingopyxis alaskensis	1.5
Staphylococcus aureus	0.8
Staphylococcus aureus	1.0
Staphylococcus aureus	1.0
Staphylococcus aureus	1.0
Staphylococcus aureus	1.0
Staphylococcus aureus	0.8
Staphylococcus aureus	0.8
Staphylococcus aureus	1.0
Staphylococcus aureus	0.8
Staphylococcus epidermidis	1.0
Staphylococcus epidermidis	0.8
Staphylococcus haemolyticus	0.8
Staphylococcus saprophyticus	1.3
Streptococcus agalactiae	2.0
Streptococcus agalactiae	2.3
Streptococcus agalactiae	2.3
Streptococcus mutans	1.0
Streptococcus pneumoniae	0.8
Streptococcus pneumoniae	1.3
Streptococcus pyogenes	0.8
Streptococcus pyogenes	1.0
Streptococcus pyogenes	0.8
Streptococcus pyogenes	0.8
Streptococcus pyogenes	0.8
Streptococcus pyogenes	0.8
Streptococcus pyogenes	0.8
Streptococcus pyogenes	1.0
Streptococcus pyogenes	0.8
Streptococcus pyogenes	0.8

(*Continued*)

Known as:	Idaho Bacteria Average
Streptomyces avermitilis	3.3
Streptomyces coelicolor	3.8
Sulfolobus acidocaldarius	0.5
Sulfolobus solfataricus	0.8
Sulfolobus tokodaii	0.8
Symbiobacterium thermophilum	1.5
Synechococcus	1.3
Synechococcus elongatus	0.8
Synechococcus elongatus	0.8
Synechococcus sp.	0.8
Synechococcus sp.	1.5
Synechococcus sp.	1.3
Synechococcus sp.	1.0
Synechocystis	0.8
Syntrophomonas wolfei	0.5
Syntrophus aciditrophicus	1.5
Thermoanaerobacter tengcongensis	0.5
Thermobifida fusca	0.8
Thermococcus kodakarensis	0.8
Thermoplasma acidophilum	1.3
Thermoplasma volcanium	1.3
Thermosynechococcus elongatus	0.8
Thermotoga maritima	1.8
Thermus thermophilus	1.5
Thermus thermophilus	1.3
Thiobacillus denitrificans	1.0
Thiomicrospira denitrificans	0.5
Treponema denticola	2.0
Trichodesmium erythraeum	1.8
Tropheryma whipplei	1.3
Tropheryma whipplei	1.3
Vibrio cholerae O1 biovar eltor	1.0
Vibrio fischeri	3.0
Vibrio parahaemolyticus	1.3
Vibrio vulnificus	1.0
Vibrio vulnificus	1.8
Wolbachia endosymbiont of *Drosophila melanogaster*	0.5
Wolinella succinogenes	1.8
Xanthomonas axonopodis pv. citri	0.5
Xanthomonas campestris pv. campestris	1.3

(*Continued*)

Known as:	Idaho Bacteria Average
Xanthomonas campestris pv. campestris str.	1.0
Xanthomonas campestris pv. vesicatoria str.	1.5
Xanthomonas oryzae	1.3
Xanthomonas oryzae pv. oryzae	1.5
Xylella fastidiosa	0.8
Zymomonas mobilis	0.5

Appendix E – Iowa Region Average for Bacteria

Known as:	Iowa Bacteria Average
Aquifex aeolicus	1.0
Acidobacteria bacterium	4.1
Acinetobacter	2.4
Aeropyrum pernix	2.0
Agrobacterium tumefaciens	6.4
Alcanivorax borkumensis	2.5
Alkalilimnicola ehrlichei	2.6
Anabaena variabilis	5.3
Anaeromyxobacter dehalogenans	5.5
Anaplasma marginale	1.1
Anaplasma phagocytophilum	1.3
Archaeoglobus fulgidus	2.0
Azoarcus	3.9
Bacillus anthracis	4.5
Bacillus anthracis	4.7
Bacillus anthracis	4.8
Bacillus cereus	4.8
Bacillus cereus	4.9
Bacillus cereus	5.0
Bacillus clausii	3.0
Bacillus halodurans	2.2
Bacillus licheniformis	3.6
Bacillus subtilis	3.8
Bacillus thuringensis serovar konkukian	4.5
Bacillus thuringiensis	4.3
Bacteroides fragilis	4.9
Bacteroides fragilis	4.8
Bacteroides thetaiotaomicron	8.2
Bartonella henselae	1.6
Bartonella quintana	1.4

(*Continued*)

Known as:	Iowa Bacteria Average
Baumannia cicadellinicola	0.4
Bdellovibrio bacteriovorus	3.7
Bifidobacterium longum	2.0
Bordetella bronchiseptica	5.2
Bordetella parapertussis	4.3
Bordetella pertussis	3.5
Borrelia afzelii	1.3
Borrelia burgdorferi	1.4
Borrelia garinii	1.5
Bradyrhizobium japonicum	7.9
Brucella abortus biovar 1	2.8
Brucella melitensis	3.2
Brucella melitensis biovar Abortus	2.9
Brucella suis	3.2
Buchnera aphidicola	0.6
Buchnera aphidicola	0.8
Buchnera aphidicola	1.1
Burkholderia cenocepacia	5.6
Burkholderia mallei	5.4
Burkholderia pseudomallei	9.0
Burkholderia pseudomallei	6.2
Burkholderia sp.	6.3
Burkholderia thailandensis	6.2
Burkholderia xenovorans	9.0
Campylobacter jejuni	1.7
Campylobacter jejuni	1.5
Candidatus Blochmannia floridanus	0.4
Candidatus Blochmannia pennsylvanicus str.	0.6
Candidatus Pelagibacter ubique	1.4
Candidatus Protochlamydia amoebophila	2.2
Carboxydothermus hydrogenoformans	2.7
Caulobacter crescentus	2.9
Chlamydia muridarum	1.0
Chlamydia trachomatis	1.0
Chlamydia trachomatis	1.0
Chlamydophila abortus	1.1
Chlamydophila caviae	0.9
Chlamydophila felis	0.8
Chlamydophila pneumoniae	0.6
Chlamydophila pneumoniae	0.6
Chlamydophila pneumoniae	0.6

(*Continued*)

Known as:	Iowa Bacteria Average
Chlamydophila pneumoniae	0.6
Chlorobium chlorochromatii	1.8
Chlorobium tepidum	3.0
Chromobacterium violaceum	5.1
Chromohalobacter salexigens	2.5
Clostridium acetobutylicum	3.5
Clostridium perfringens	3.0
Clostridium perfringens	3.4
Clostridium perfringens	2.6
Clostridium tetani	3.5
Colwellia psychrerythraea	3.9
Corynebacterium diphtheriae	1.6
Corynebacterium efficiens	2.5
Corynebacterium glutamicum	2.8
Corynebacterium jeikeium	2.0
Coxiella burnetii	1.4
Cytophaga hutchinsonii	3.8
Dechloromonas aromatica	4.3
Dehalococcoides ethenogenes	6.1
Dehalococcoides sp.	2.0
Deinococcus geothermalis	3.1
Deinococcus radiodurans	2.8
Desulfitobacterium hafniense	4.8
Desulfotalea psychrophila	2.7
Desulfovibrio desulfuricans	3.2
Desulfovibrio vulgaris	3.7
Ehrlichia canis str.	1.1
Ehrlichia chaffeensis str.	0.8
Ehrlichia ruminantium	0.7
Ehrlichia ruminantium	1.0
Enterococcus faecalis	2.4
Erwinia carotovoa	3.2
Erythrobacter litoralis	3.1
Escherichia coli	4.3
Escherichia coli	4.2
Escherichia coli	4.8
Escherichia coli	4.1
Escherichia coli	4.6
Escherichia coli	4.6
Escherichia coli	4.9
Escherichia coli	4.1
Fowlpox Virus	0.2

(*Continued*)

Known as:	Iowa Bacteria Average
Francisella tularensis	1.0
Francisella tularensis	1.1
Francisella tularensis	1.0
Francisella tularensis	1.0
Frankia alni	6.3
Frankia sp.	3.8
Fusobacterium nucleatum	3.0
Geobacillus kausophilus	3.9
Geobacter metallireducens	4.2
Geobacter sulfurreducens	3.4
Gloeobacter violaceus	3.8
Gluconobacter oxydans	2.8
Granulibacter bethesdensis	2.2
Haemophilus ducreyi	1.4
Haemophilus influenzae	1.6
Haemophilus influenzae	1.3
Haemophilus somnus	1.7
Hahella chejuensis	6.0
Haloarcula marismortui	2.0
Halobacterium	1.6
Haloquadratum walsbyi	1.4
Helicobacter acinonychis str.	1.5
Helicobacter hepaticus	1.7
Helicobacter pylori	1.8
Helicobacter pylori	1.5
Helicobacter pylori	1.9
Hyphomonas neptunium	3.6
Idiomarina loihiensis	2.2
Jannaschia sp.	3.3
Klebsiella pneumoniae	4.4
Lactobacillus acidophilus	1.4
Lactobacillus delbrueckii	1.6
Lactobacillus johnsonii	1.6
Lactobacillus plantarum	2.8
Lactobacillus sakei	1.3
Lactobacillus salivarius	1.8
Lactococcus lactis	2.0
Lawsonia intracellularis	1.3
Legionella pneumophila	2.2
Legionella pneumophila	2.2
Legionella pneumophila	2.4

(*Continued*)

Known as:	Iowa Bacteria Average
Leifsonia xyli	1.1
Leptospira interrogans serovar Copenhageni	3.9
Leptospira interrogans serovar lai	4.0
Listeria innocua	2.7
Listeria monocytogenes	1.8
Listeria monocytogenes	2.1
Magnetospirillum magneticum	4.7
Mannheimia succiniproducens	1.7
Maricaulis maris	3.0
Mesoplasma florum	0.7
Mesorhizobium loti	6.9
Mesorhizobium sp.	4.0
Methanocaldococcus jannaschii	1.8
Methanococcoides burtonii	2.0
Methanococcus maripaludis	2.0
Methanopyrus kandleri	1.6
Methanosarcina acetivorans	4.1
Methanosarcina barkeri str.	3.4
Methanosarcina mazei	2.5
Methanosphaera stadtmanae	1.8
Methanospirillum hungatei	2.7
Methanothermobacter thermautotrophicus	1.4
Methylobacillus flagellatus	2.9
Methylococcus capsulatus	2.6
Moorella thermoacetica	2.9
Mycobacterium avium	4.2
Mycobacterium bovis	3.4
Mycobacterium leprae	1.7
Mycobacterium sp.	5.2
Mycobacterium tuberculosis	3.8
Mycobacterium tuberculosis	3.6
Mycoplasma capricolum	1.1
Mycoplasma gallisepticum	0.8
Mycoplasma genitalium	0.6
Mycoplasma hyopneumoniae	0.8
Mycoplasma hyopneumoniae	0.6
Mycoplasma hyopneumoniae	0.8
Mycoplasma mobile	0.7
Mycoplasma mycoides	1.2
Mycoplasma penetrans	0.8

(*Continued*)

Known as:	Iowa Bacteria Average
Mycoplasma pneumoniae	0.8
Mycoplasma pulmonis	1.0
Mycoplasma synoviae	0.7
Myxococcus xanthus	10.0
Nanoarchaeum equitans	0.4
Natronomonas pharaonis	1.4
Neisseria gonorrhoeae	1.8
Neisseria meningitidis	2.0
Neisseria meningitidis	2.0
Neorickettsia sennetsu str.	1.1
Nitrobacter hamburgensis	5.2
Nitrobacter winogradskyi	2.9
Nitrosococcus oceani	2.6
Nitrosomonas europaea	2.4
Nitrosomonas eutropha	2.3
Nitrosospira multiformis	2.7
Nocardia farcinica	6.2
Nostoc sp.	5.6
Novosphingobium aromaticivorans	3.8
Oceanobacillus iheyensis	2.4
Onion Yellows Phytoplasma	0.7
Pasteurella Multocida	2.2
Pelobacter carbinolicus	2.6
Pelodictyon luteolum	2.3
Photobacterium profundum	4.6
Photorhabdus luminescens	3.7
Picrophilus torridus	1.3
Polaromonas sp.	4.5
Porphyromonas gingivalis	1.9
Prochlorococcus marinus	1.5
Prochlorococcus marinus	2.0
Prochlorococcus marinus	1.7
Prochlorococcus marinus str.	1.2
Prochlorococcus marinus str.	1.5
Propionibacterium acnes	2.0
Pseudoalteromonas atlantica	3.8
Pseudoalteromonas haloplanktis	3.4
Pseudomonas aeruginosa	5.0
Pseudomonas entomophila	5.2
Pseudomonas fluorescens	5.2
Pseudomonas fluorescens	6.2
Pseudomonas putida	5.5

(*Continued*)

Known as:	**Iowa Bacteria Average**
Pseudomonas syringae pv. phaseolicola	5.5
Pseudomonas syringae pv. syringae	4.7
Pseudomonas syringae pv. tomato	5.4
Psychrobacter arcticus	1.6
Psychrobacter cryohalolentis	2.0
Pyrobaculum aerophilum	2.3
Pyrococcus abyssi	2.0
Pyrococcus furiosus	1.8
Pyrococcus horikoshii	2.4
Ralstonia eutropha	4.0
Ralstonia eutropha	6.9
Ralstonia metallidurans	6.4
Ralstonia solanacearum	4.6
Rhizobium etli	6.1
Rhodobacter sphaeroides	5.1
Rhodococcus sp.	8.6
Rhodoferax ferrireducens	4.2
Rhodopirellula baltica	5.5
Rhodopseudomonas palustris	5.0
Rhodopseudomonas palustris	5.2
Rhodopseudomonas palustris	5.7
Rhodopseudomonas palustris	5.3
Rhodospirillum rubrum	4.5
Rickettsia bellii	1.3
Rickettsia conorii	1.0
Rickettsia felis	1.4
Rickettsia prowazekii	1.0
Rickettsia typhi	0.6
Roseobacter denitrificans	3.4
Rubrobacter xylanophilus	3.4
Saccharophagus degradans	3.7
Salinibacter ruber	2.4
Salmonella enterica	3.8
Salmonella enterica	4.0
Salmonella enterica	4.2
Salmonella enterica	3.8
Salmonella typhimurium	5.0
Shewanella denitrificans	3.8
Shewanella frigidimarina	2.8
Shewanella oneidensis	3.0
Shewanella sp.	3.2
Shewanella sp.	3.0

(*Continued*)

Known as:	Iowa Bacteria Average
Shigella boydii	3.6
Shigella dysenteriae	3.2
Shigella flexneri 2a	3.3
Shigella flexneri 2a	3.5
Shigella flexneri 5 str.	3.2
Shigella sonnei	3.7
Silicibacter pomeroyi	4.5
Silicibacter sp.	3.8
Sinorhizobium meliloti	5.6
Sodalis glossinidius str.	2.1
Sphingopyxis alaskensis	3.1
Staphylococcus aureus	2.8
Staphylococcus aureus	2.7
Staphylococcus aureus	2.6
Staphylococcus aureus	2.7
Staphylococcus aureus	2.7
Staphylococcus aureus	2.6
Staphylococcus aureus	2.6
Staphylococcus aureus	2.1
Staphylococcus aureus	2.5
Staphylococcus epidermidis	2.8
Staphylococcus epidermidis	2.1
Staphylococcus haemolyticus	2.2
Staphylococcus saprophyticus	2.3
Staphylococcus thermophilus	1.6
Streptococcus agalactiae	2.1
Streptococcus agalactiae	2.2
Streptococcus agalactiae	2.0
Streptococcus mutans	1.9
Streptococcus pneumoniae	1.5
Streptococcus pneumoniae	1.6
Streptococcus pyogenes	1.6
Streptococcus pyogenes	1.8
Streptococcus pyogenes	1.5
Streptococcus pyogenes	1.8
Streptococcus pyogenes	1.6
Streptococcus pyogenes	1.6
Streptococcus pyogenes	1.5
Streptococcus pyogenes	1.6
Streptococcus pyogenes	1.6
Streptococcus pyogenes	1.6
Streptococcus pyogenes	1.6

(*Continued*)

Known as:	Iowa Bacteria Average
Streptococcus thermophilus	1.7
Streptomyces avermitilis	9.3
Streptomyces coelicolor	9.0
Sulfolobus acidocaldarius	2.4
Sulfolobus solfataricus	2.4
Sulfolobus tokodaii	2.2
Symbiobacterium thermophilum	3.0
Synechococcus	1.6
Synechococcus elongatus	1.8
Synechococcus elongatus	1.7
Synechococcus sp.	2.0
Synechococcus sp.	2.1
Synechococcus sp.	1.8
Synechococcus sp.	2.4
Synechococcus sp.	2.0
Synechocystis	2.5
Syntrophomonas wolfei	2.2
Syntrophus aciditrophicus	3.3
Thermoanaerobacter tengcongensis	3.0
Thermobifida fusca	2.6
Thermococcus kodakarensis	1.8
Thermoplasma acidophilum	1.1
Thermoplasma volcanium	1.5
Thermosynechococcus elongatus	2.0
Thermotoga maritima	1.8
Thermus thermophilus	2.2
Thermus thermophilus	2.4
Thiobacillus denitrificans	3.5
Thiomicrospira crunogena	2.0
Thiomicrospira denitrificans	2.1
Treponema denticola	2.1
Treponema pallidum	1.5
Trichodesmium erythraeum	3.9
Tropheryma whipplei	1.0
Tropheryma whipplei	1.0
Ureaplasma urealyticum	0.5
Vibrio cholerae O1 biovar eltor	3.0
Vibrio fischeri	3.5
Vibrio parahaemolyticus	3.2
Vibrio vulnificus	3.3

(*Continued*)

Known as:	Iowa Bacteria Average
Vibrio vulnificus	3.8
Wigglesworthia glossinidia endosymbiont of Glossina brevipalpis	1.0
Wolbachia endosymbiont	0.8
Wolbachia endosymbiont of Drosophila melanogaster	1.0
Wolinella succinogenes	1.9
Xanthomonas axonopodis pv. citri	3.6
Xanthomonas campestris pv. campestris	4.4
Xanthomonas campestris pv. campestris str.	5.1
Xanthomonas campestris pv. vesicatoria str.	3.6
Xanthomonas oryzae	3.2
Xanthomonas oryzae pv. oryzae	3.2
Xylella fastidiosa	2.1
Xylella fastidiosa	2.0
Yersenia pseudotuberculosis	3.3
Yersinia pestis	3.2
Yersinia pestis	3.1
Yersinia pestis	3.2
Yersinia pestis	3.0
Yersinia pestis biovar Mediaevails	3.1
Zymomonas mobilis	1.5

Appendix F – Montana Regional Average for Bacteria

Listing of bacteria detected and identified along with the average number of unique peptides.

Known as	Montana Bacteria Average
Acidobacteria bacterium	15.5
Acinetobacter	4.5
Aeropyrum pernix	4.0
Agrobacterium tumefaciens	7.5
Alcanivorax borkumensis	5.0
Alkalilimnicola ehrlichei	3.5
Anabaena variabilis	9.5
Anaeromyxobacter dehalogenans	13.5
Anaplasma marginale	2.5
Anaplasma phagocytophilum	0.6
Aquifex aeolicus	5.5
Archaeoglobus fulgidus	3.0
Azoarcus	10.5
Bacillus anthracis	6.5
Bacillus anthracis	7.0
Bacillus anthracis	7.0
Bacillus cereus	6.0
Bacillus cereus	9.5
Bacillus cereus	7.0
Bacillus clausii	8.5
Bacillus halodurans	5.0
Bacillus licheniformis	5.5
Bacillus subtilis	10.0
Bacillus thuringensis serovar konkukian	7.0
Bacillus thuringiensis	7.5
Bacteroides fragilis	10.0
Bacteroides fragilis	10.0
Bacteroides thetaiotaomicron	14.5

(*Continued*)

Known as	Montana Bacteria Average
Bartonella henselae	2.5
Bartonella quintana	1.5
Baumannia cicadellinicola	1.0
Bdellovibrio bacteriovorus	6.0
Bifidobacterium longum	3.0
Bordetella bronchiseptica	8.5
Bordetella parapertussis	7.0
Bordetella pertussis	6.5
Borrelia afzelii	2.0
Borrelia burgdorferi	2.0
Borrelia garinii	3.0
Bradyrhizobium japonicum	14.5
Brucella abortus biovar 1	4.5
Brucella melitensis	5.0
Brucella melitensis biovar Abortus	5.0
Brucella suis	4.0
Buchnera aphidicola	1.0
Buchnera aphidicola	0.5
Burkholderia cenocepacia	9.0
Burkholderia mallei	12.0
Burkholderia pseudomallei	18.0
Burkholderia pseudomallei	12.5
Burkholderia sp.	11.5
Burkholderia thailandensis	13.0
Burkholderia xenovorans	14.5
Campylobacter jejuni	1.0
Campylobacter jejuni	2.0
Candidatus Blochmannia floridanus	0.5
Candidatus Blochmannia pennsylvanicus str.	1.0
Candidatus Pelagibacter ubique	2.5
Candidatus Protochlamydia amoebophila	2.0
Carboxydothermus hydrogenoformans	4.0
Caulobacter crescentus	8.5
Chlamydia muridarum	1.5
Chlamydophila abortus	1.5
Chlamydophila felis	1.5
Chlamydophila pneumoniae	1.5
Chlamydophila pneumoniae	1.5
Chlamydophila pneumoniae	1.5
Chlamydophila pneumoniae	1.5
Chlorobium chlorochromatii	2.0

(*Continued*)

Known as	Montana Bacteria Average
Chlorobium tepidum	2.5
Chromobacterium violaceum	4.5
Chromohalobacter salexigens	7.0
Clostridium acetobutylicum	7.5
Clostridium perfringens	9.5
Clostridium perfringens	8.5
Clostridium perfringens	7.5
Clostridium tetani	5.0
Colwellia psychrerythraea	6.0
Corynebacterium diphtheriae	2.5
Corynebacterium efficiens	3.0
Corynebacterium glutamicum	3.5
Corynebacterium jeikeium	4.0
Coxiella burnetii	3.0
Cytophaga hutchinsonii	6.0
Dechloromonas aromatica	7.0
Dehalococcoides ethenogenes	12.0
Dehalococcoides sp.	2.0
Deinococcus geothermalis	6.0
Deinococcus radiodurans	8.5
Desulfitobacterium hafniense	10.0
Desulfotalea psychrophila	6.0
Desulfovibrio desulfuricans	11.5
Desulfovibrio vulgaris	5.0
Ehrlichia canis str.	2.5
Ehrlichia chaffeensis str.	2.0
Ehrlichia ruminantium	2.5
Ehrlichia ruminantium	2.5
Enterococcus faecalis	4.0
Erwinia carotovoa	6.5
Erythrobacter litoralis	5.5
Escherichia coli	9.0
Escherichia coli	10.0
Escherichia coli	10.0
Escherichia coli	7.5
Escherichia coli	9.5
Escherichia coli	10.0
Escherichia coli	11.0
Escherichia coli	7.5
Fowlpox Virus	0.2
Francisella tularensis	1.0
Francisella tularensis	1.5

(*Continued*)

Known as	Montana Bacteria Average
Francisella tularensis	1.5
Francisella tularensis	1.0
Frankia alni	14.5
Frankia sp.	9.0
Fusobacterium nucleatum	3.5
Geobacillus kausophilus	6.0
Geobacter metallireducens	5.5
Geobacter sulfurreducens	3.0
Gloeobacter violaceus	10.0
Gluconobacter oxydans	3.0
Granulibacter bethesdensis	4.0
Haemophilus ducreyi	1.5
Haemophilus influenzae	1.0
Haemophilus influenzae	1.0
Haemophilus somnus	3.0
Hahella chejuensis	10.0
Haloarcula marismortui	5.0
Halobacterium	2.0
Haloquadratum walsbyi	4.0
Helicobacter acinonychis str.	1.0
Helicobacter hepaticus	1.5
Helicobacter pylori	3.5
Helicobacter pylori	1.0
Helicobacter pylori	0.5
Hyphomonas neptunium	8.0
Idiomarina loihiensis	4.5
Jannaschia sp.	5.5
Klebsiella pneumoniae	8.0
Lactobacillus acidophilus	6.0
Lactobacillus delbrueckii	2.5
Lactobacillus johnsonii	3.0
Lactobacillus plantarum	2.5
Lactobacillus sakei	2.5
Lactobacillus salivarius	2.5
Lactococcus lactis	6.0
Lawsonia intracellularis	1.5
Legionella pneumophila	8.0
Legionella pneumophila	5.5
Legionella pneumophila	8.0
Leifsonia xyli	4.0
Leptospira interrogans serovar Copenhageni	5.0

(*Continued*)

Known as	Montana Bacteria Average
Leptospira interrogans serovar lai	4.5
Listeria innocua	3.5
Listeria monocytogenes	3.5
Listeria monocytogenes	5.0
Magnetospirillum magneticum	8.0
Mannheimia succiniproducens	4.5
Maricaulis maris	4.5
Mesoplasma florum	2.0
Mesorhizobium loti	10.5
Mesorhizobium sp.	8.5
Methanocaldococcus jannaschii	3.5
Methanococcoides burtonii	2.5
Methanococcus maripaludis	2.5
Methanopyrus kandleri	1.5
Methanosarcina acetivorans	8.0
Methanosarcina barkeri str.	6.5
Methanosarcina mazei	7.0
Methanosphaera stadtmanae	2.5
Methanospirillum hungatei	7.0
Methanothermobacter thermautotrophicus	2.0
Methylobacillus flagellatus	4.5
Methylococcus capsulatus	8.0
Moorella thermoacetica	3.0
Mycobacterium avium	9.0
Mycobacterium bovis	8.5
Mycobacterium leprae	3.0
Mycobacterium sp.	12.0
Mycobacterium tuberculosis	10.5
Mycobacterium tuberculosis	9.0
Mycoplasma capricolum	0.5
Mycoplasma gallisepticum	1.5
Mycoplasma genitalium	1.0
Mycoplasma hyopneumoniae	2.5
Mycoplasma hyopneumoniae	2.5
Mycoplasma hyopneumoniae	2.0
Mycoplasma mobile	1.5
Mycoplasma mycoides	3.0
Mycoplasma penetrans	2.5
Mycoplasma pneumoniae	2.0
Mycoplasma pulmonis	1.0
Mycoplasma synoviae	0.3

(*Continued*)

Known as	Montana Bacteria Average
Myxococcus xanthus	18.5
Nanoarchaeum equitans	1.0
Natronomonas pharaonis	3.0
Neisseria gonorrhoeae	4.5
Neisseria meningitidis	4.0
Neisseria meningitidis	4.5
Neorickettsia sennetsu str.	1.0
Nitrobacter hamburgensis	5.0
Nitrobacter winogradskyi	7.5
Nitrosococcus oceani	7.0
Nitrosomonas europaea	3.0
Nitrosomonas eutropha	1.0
Nitrosospira multiformis	4.0
Nocardia farcinica	11.0
Nostoc sp.	8.0
Novosphingobium aromaticivorans	5.5
Oceanobacillus iheyensis	5.0
Onion Yellows Phytoplasma	1.5
Pasteurella Multocida	2.0
Pelobacter carbinolicus	3.5
Pelodictyon luteolum	3.5
Photobacterium profundum	5.5
Photorhabdus luminescens	9.0
Picrophilus torridus	3.0
Polaromonas sp.	12.5
Porphyromonas gingivalis	2.5
Prochlorococcus marinus	1.5
Prochlorococcus marinus	2.5
Prochlorococcus marinus	5.0
Prochlorococcus marinus str.	4.0
Prochlorococcus marinus str.	4.5
Propionibacterium acnes	3.0
Pseudoalteromonas atlantica	11.0
Pseudoalteromonas haloplanktis	3.0
Pseudomonas aeruginosa	8.5
Pseudomonas entomophila	2.5
Pseudomonas fluorescens	10.5
Pseudomonas fluorescens	11.0
Pseudomonas putida	7.5
Pseudomonas syringae pv. phaseolicola	9.0
Pseudomonas syringae pv. syringae	11.0
Pseudomonas syringae pv. tomato	11.0

(Continued)

Known as	Montana Bacteria Average
Psychrobacter arcticus	2.5
Psychrobacter cryohalolentis	3.5
Pyrobaculum aerophilum	1.5
Pyrococcus abyssi	4.5
Pyrococcus furiosus	3.0
Pyrococcus horikoshii	5.0
Ralstonia eutropha	7.5
Ralstonia eutropha	14.5
Ralstonia metallidurans	13.5
Ralstonia solanacearum	8.5
Rhizobium etli	9.5
Rhodobacter sphaeroides	9.0
Rhodococcus sp.	13.0
Rhodoferax ferrireducens	5.5
Rhodopirellula baltica	9.5
Rhodopseudomonas palustris	10.5
Rhodopseudomonas palustris	6.5
Rhodopseudomonas palustris	7.0
Rhodopseudomonas palustris	8.5
Rhodospirillum rubrum	2.5
Rickettsia bellii	0.5
Rickettsia conorii	1.0
Rickettsia felis	2.0
Rickettsia prowazekii	2.0
Rickettsia typhi	2.0
Roseobacter denitrificans	4.5
Rubrobacter xylanophilus	5.0
Saccharophagus degradans	11.0
Salinibacter ruber	6.0
Salmonella enterica	6.0
Salmonella enterica	5.5
Salmonella enterica	6.5
Salmonella enterica	5.5
Salmonella typhimurium	6.0
Shewanella denitrificans	6.0
Shewanella frigidimarina	4.5
Shewanella oneidensis	6.5
Shewanella sp.	6.5
Shewanella sp.	6.0
Shigella boydii	6.5
Shigella dysenteriae	7.0
Shigella flexneri 2a	6.5

(*Continued*)

Known as	Montana Bacteria Average
Shigella flexneri 2a	7.5
Shigella flexneri 5 str.	7.5
Shigella sonnei	6.5
Silicibacter pomeroyi	6.5
Silicibacter sp.	4.5
Sinorhizobium meliloti	7.5
Sodalis glossinidius str.	4.0
Sphingopyxis alaskensis	7.0
Staphylococcus aureus	3.5
Staphylococcus aureus	3.5
Staphylococcus aureus	3.5
Staphylococcus aureus	4.0
Staphylococcus aureus	4.0
Staphylococcus aureus	3.5
Staphylococcus aureus	4.0
Staphylococcus aureus	4.0
Staphylococcus aureus	3.5
Staphylococcus epidermidis	7.0
Staphylococcus epidermidis	6.0
Staphylococcus haemolyticus	2.0
Staphylococcus saprophyticus	3.0
Staphylococcus thermophilus	5.5
Streptococcus agalactiae	1.0
Streptococcus agalactiae	0.5
Streptococcus mutans	4.5
Streptococcus pneumoniae	4.5
Streptococcus pneumoniae	4.0
Streptococcus pyogenes	4.0
Streptococcus pyogenes	3.5
Streptococcus pyogenes	3.5
Streptococcus pyogenes	4.0
Streptococcus pyogenes	4.0
Streptococcus pyogenes	4.5
Streptococcus pyogenes	3.5
Streptococcus pyogenes	5.5
Streptococcus pyogenes	3.0
Streptococcus pyogenes	4.5
Streptococcus pyogenes	3.5
Streptococcus thermophilus	5.5
Streptomyces avermitilis	10.0
Streptomyces coelicolor	16.0
Sulfolobus acidocaldarius	2.5

(*Continued*)

Known as	Montana Bacteria Average
Sulfolobus solfataricus	5.5
Sulfolobus tokodaii	2.0
Symbiobacterium thermophilum	4.5
Synechococcus	3.0
Synechococcus elongatus	5.0
Synechococcus elongatus	5.0
Synechococcus sp.	4.0
Synechococcus sp.	3.0
Synechococcus sp.	4.5
Synechococcus sp.	3.0
Synechococcus sp.	5.0
Synechocystis	5.5
Syntrophomonas wolfei	4.5
Syntrophus aciditrophicus	3.0
Thermoanaerobacter tengcongensis	7.0
Thermobifida fusca	3.5
Thermococcus kodakarensis	2.5
Thermoplasma acidophilum	2.0
Thermoplasma volcanium	2.5
Thermosynechococcus elongatus	2.0
Thermotoga maritima	3.5
Thermus thermophilus	6.0
Thermus thermophilus	7.0
Thiobacillus denitrificans	5.0
Thiomicrospira crunogena	4.5
Thiomicrospira denitrificans	2.5
Treponema denticola	5.0
Treponema pallidum	2.5
Trichodesmium erythraeum	4.0
Tropheryma whipplei	1.5
Tropheryma whipplei	1.5
Ureaplasma urealyticum	1.0
Vibrio cholerae O1 biovar eltor	5.5
Vibrio fischeri	5.5
Vibrio parahaemolyticus	10.0
Vibrio vulnificus	6.0
Vibrio vulnificus	4.5
Wigglesworthia glossinidia endosymbiont of *Glossina brevipalpis*	1.5
Wolbachia endosymbiont	2.0
Wolbachia endosymbiont *Drosophila melanogaster*	2.0

(*Continued*)

Known as	Montana Bacteria Average
Wolinella succinogenes	5.5
Xanthomonas axonopodis pv. citri	7.5
Xanthomonas campestris pv. campestris	6.5
Xanthomonas campestris pv. campestris str.	6.5
Xanthomonas campestris pv. vesicatoria str.	9.5
Xanthomonas oryzae	5.0
Xanthomonas oryzae pv. oryzae	5.0
Xylella fastidiosa	2.0
Xylella fastidiosa	3.0
Yersenia pseudotuberculosis	5.0
Yersinia pestis	6.5
Yersinia pestis	5.0
Yersinia pestis	6.0
Yersinia pestis	5.0
Yersinia pestis biovar Mediaevails	6.0
Zymomonas mobilis	0.5

Appendix G – National Average for Fungi

Listing of fungi detected and identified along with average number of unique peptides.

Known as:	National Fungi Average
[Candida] pseudohaemulonis	3.2
Agaricus bisporus var. bisporus H97	5.7
Alternaria alternata	9.7
Amorphotheca resinae ATCC 22711	6.6
Ascoidea rubescens DSM 1968	3.4
Aspergillus aculeatus ATCC 16872	6.6
Aspergillus bombycis	7.5
Aspergillus campestris IBT 28561	8.6
Aspergillus candidus	8.1
Aspergillus glaucus CBS 516.65	6.3
Aspergillus novofumigatus IBT 16806	8.0
Aspergillus oryzae RIB40	6.6
Aspergillus steynii IBT 23096	8.2
Babjeviella inositovora NRRL Y-12698	3.6
Baudoinia panamericana UAMH 10762	5.9
Bipolaris maydis ATCC 48331	7.5
Bipolaris sorokiniana ND90Pr	7.5
Botrytis cinerea B05.10	7.8
Candida albicans SC5314	3.4
Candida tropicalis MYA-3404	4.0
Cercospora beticola	8.4
Coccidioides posadasii C735 delta SOWgp	6.0
Colletotrichum gloeosporioides Nara gc5	10.3
Colletotrichum higginsianum IMI 349063	8.8
Colletotrichum orchidophilum	9.3
Cordyceps fumosorosea ARSEF 2679	7.9
Cryptococcus amylolentus CBS 6039	6.1

(*Continued*)

Known as:	National Fungi Average
Cyberlindnera jadinii NRRL Y-1542	3.5
Debaryomyces fabryi	3.7
Diplodia corticola	8.7
Eremothecium sinecaudum	3.2
es blakesleeanus NRRL 1555(-)	6.8
Eutypa lata UCREL1	7.0
Exserohilum turcica Et28A	8.4
Fonsecaea erecta	7.9
Fonsecaea monophora	8.1
Fonsecaea nubica	7.6
Fusarium fujikuroi IMI 58289	8.6
Gloeophyllum trabeum ATCC 11539	7.4
Hyphopichia burtonii NRRL Y-1933	4.0
Kazachstania naganishii CBS 8797	3.8
Kluyveromyces marxianus DMKU3–1042	3.6
Kockovaella imperatae	5.5
Kwoniella bestiolae CBS 10118	6.7
Kwoniella dejecticola CBS 10117	6.7
Kwoniella mangroviensis CBS 8507	6.3
Kwoniella pini CBS 10737	5.4
Lobosporangium transversale	7.8
Meliniomyces bicolor E	10.7
Metschnikowia bicuspidata var. bicuspidata NRRL YB-4993	3.4
Millerozyma farinosa CBS 7064	5.0
Neofusicoccum parvum UCRNP2	6.7
Ogataea polymorpha	2.9
Paraphaeosphaeria sporulosa	9.3
Penicilliopsis zonata CBS 506.65	6.8
Penicillium arizonense	7.3
Penicillium digitatum Pd1	5.9
Phaeoacremonium minimum UCRPA7	6.6
Phialocephala scopiformis	10.1
Pichia kudriavzevii	2.8
Pichia membranifaciens NRRL Y-2026	3.6
Pneumocystis carinii B80	2.3
Pneumocystis jirovecii RU7	3.0
Pneumocystis murina B123	2.7
Pochonia chlamydosporia 170	9.4
Postia placenta MAD-698-R-SB12	7.1
Pseudocercospora fijiensis CIRAD86	8.2
Pseudogymnoascus destructans	6.8

(*Continued*)

Known as:	National Fungi Average
Pseudogymnoascus destructans 20631-21	7.0
Pseudogymnoascus verrucosus	7.8
Pseudozyma hubeiensis SY62	6.8
Purpureocillium lilacinum	9.2
Ramularia collo-cygni	7.5
Rhizopus microsporus ATCC 52813	5.3
Rhodotorula graminis WP1	6.2
Rhodotorula toruloides NP11	6.3
Saccharomyces eubayanus	3.8
Saitoella complicata NRRL Y-17804	4.5
Sclerotinia sclerotiorum 1980 UF-70	8.0
Sphaerulina musiva SO2202	7.6
Sugiyamaella lignohabitans	3.6
Suhomyces tanzawaensis NRRL Y-17324	3.3
Talaromyces atroroseus	7.0
Tetrapisispora blattae CBS 6284	3.7
Trichoderma asperellum CBS 433.97	7.8
Trichoderma citrinoviride	7.2
Trichoderma gamsii	7.1
Trichoderma harzianum CBS 226.95	8.4
Trichosporon asahii var. asahii CBS 2479	5.7
Tsuchiyaea wingfieldii CBS 7118	6.0
Wallemia ichthyophaga EXF-994	3.6
Wickerhamiella sorbophila	3.0
Wickerhamomyces anomalus NRRL Y-366-8	4.2
Xylona heveae TC161	6.0

Appendix H – California Region Average for Fungi

Listing of Fungi detected and identified along with average number of unique peptides.

Known as:	**California Fungi Average**
[Candida] pseudohaemulonis	3.7
Agaricus bisporus var. bisporus H97	5.6
Alternaria alternata	7.9
Amorphotheca resinae ATCC 22711	6.8
Ascoidea rubescens DSM 1968	2.9
Aspergillus aculeatus ATCC 16872	5.2
Aspergillus bombycis	6.6
Aspergillus campestris IBT 28561	5.3
Aspergillus candidus	5.4
Aspergillus glaucus CBS 516.65	7.0
Aspergillus novofumigatus IBT 16806	6.6
Aspergillus oryzae RIB40	5.1
Aspergillus steynii IBT 23096	8.3
Babjeviella inositovora NRRL Y-12698	4.0
Baudoinia panamericana UAMH 10762	5.4
Bipolaris maydis ATCC 48331	6.8
Bipolaris sorokiniana ND90Pr	6.9
Botrytis cinerea B05.10	6.4
Candida albicans SC5314	3.9
Candida tropicalis MYA-3404	3.7
Cercospora beticola	7.2
Coccidioides posadasii C735 delta SOWgp	7.3
Colletotrichum gloeosporioides Nara gc5	7.9
Colletotrichum higginsianum IMI 349063	8.6
Colletotrichum orchidophilum	7.4
Cordyceps fumosorosea ARSEF 2679	6.1

(*Continued*)

Known as:	California Fungi Average
Cryptococcus amylolentus CBS 6039	5.6
Cyberlindnera jadinii NRRL Y-1542	4.1
Debaryomyces fabryi	3.1
Diplodia corticola	7.8
Eremothecium sinecaudum	3.2
es blakesleeanus NRRL 1555(-)	5.6
Eutypa lata UCRELI	6.3
Exserohilum turcica Et28A	7.8
Fonsecaea erecta	8.1
Fonsecaea monophora	7.5
Fonsecaea nubica	6.9
Fusarium fujikuroi IMI 58289	7.6
Gloeophyllum trabeum ATCC 11539	6.5
Hyphopichia burtonii NRRL Y-1933	3.4
Kazachstania naganishii CBS 8797	3.3
Kluyveromyces marxianus DMKU3–1042	3.9
Kockovaella imperatae	5.7
Kwoniella bestiolae CBS 10118	6.6
Kwoniella dejecticola CBS 10117	5.6
Kwoniella mangroviensis CBS 8507	6.4
Kwoniella pini CBS 10737	5.0
Lobosporangium transversale	6.2
Meliniomyces bicolor E	8.6
Metschnikowia bicuspidata var. bicuspidata NRRL YB-4993	2.8
Millerozyma farinosa CBS 7064	3.3
Neofusicoccum parvum UCRNP2	5.9
Ogataea polymorpha	3.6
Paraphaeosphaeria sporulosa	7.0
Penicilliopsis zonata CBS 506.65	5.4
Penicillium arizonense	6.3
Penicillium digitatum Pd1	4.3
Phaeoacremonium minimum UCRPA7	5.8
Phialocephala scopiformis	8.9
Pichia kudriavzevii	2.9
Pichia membranifaciens NRRL Y-2026	3.5
Pneumocystis carinii B80	2.4
Pneumocystis jirovecii RU7	2.3
Pneumocystis murina B123	1.9
Pochonia chlamydosporia 170	7.9
Postia placenta MAD-698-R-SB12	5.9
Pseudocercospora fijiensis CIRAD86	7.1

(*Continued*)

Known as:	California Fungi Average
Pseudogymnoascus destructans	6.2
Pseudogymnoascus destructans 20631-21	6.6
Pseudogymnoascus verrucosus	6.9
Pseudozyma hubeiensis SY62	4.6
Purpureocillium lilacinum	7.8
Ramularia collo-cygni	7.4
Rhizopus microsporus ATCC 52813	4.6
Rhodotorula graminis WP1	5.6
Rhodotorula toruloides NP11	6.4
Saccharomyces eubayanus	4.1
Saitoella complicata NRRL Y-17804	4.8
Sclerotinia sclerotiorum 1980 UF-70	5.8
Sphaerulina musiva SO2202	5.0
Sugiyamaella lignohabitans	3.2
Suhomyces tanzawaensis NRRL Y-17324	2.9
Talaromyces atroroseus	5.8
Tetrapisispora blattae CBS 6284	3.9
Trichoderma asperellum CBS 433.97	5.6
Trichoderma citrinoviride	5.4
Trichoderma gamsii	4.2
Trichoderma harzianum CBS 226.95	7.1
Trichosporon asahii var. asahii CBS 2479	6.0
Tsuchiyaea wingfieldii CBS 7118	5.1
Wallemia ichthyophaga EXF-994	3.1
Wickerhamiella sorbophila	2.5
Wickerhamomyces anomalus NRRL Y-366-8	3.5
Xylona heveae TC161	6.4

Appendix I – Florida Region Average for Fungi

Listing of fungi detected and identified along with average number of unique peptides.

Known as:	Florida Fungi Average
[Candida] pseudohaemulonis	1.4
Agaricus bisporus var. bisporus H97	2.0
Alternaria alternata	3.4
Amorphotheca resinae ATCC 22711	2.8
Ascoidea rubescens DSM 1968	1.5
Aspergillus aculeatus ATCC 16872	2.5
Aspergillus bombycis	2.6
Aspergillus campestris IBT 28561	2.5
Aspergillus candidus	2.6
Aspergillus glaucus CBS 516.65	2.7
Aspergillus novofumigatus IBT 16806	2.8
Aspergillus oryzae RIB40	3.1
Aspergillus steynii IBT 23096	3.6
Babjeviella inositovora NRRL Y-12698	1.5
Baudoinia panamericana UAMH 10762	2.6
Bipolaris maydis ATCC 48331	3.2
Bipolaris sorokiniana ND90Pr	3.2
Botrytis cinerea B05.10	3.2
Candida albicans SC5314	1.3
Candida tropicalis MYA-3404	1.2
Cercospora beticola	3.2
Coccidioides posadasii C735 delta SOWgp	2.4
Colletotrichum gloeosporioides Nara gc5	4.6
Colletotrichum higginsianum IMI 349063	4.1
Colletotrichum orchidophilum	3.2
Cordyceps fumosorosea ARSEF 2679	3.0
Cryptococcus amylolentus CBS 6039	2.8

(*Continued*)

Known as:	Florida Fungi Average
Cyberlindnera jadinii NRRL Y-1542	1.5
Debaryomyces fabryi	1.5
Diplodia corticola	3.2
Eremothecium sinecaudum	1.3
es blakesleeanus NRRL 1555(-)	2.7
Eutypa lata UCREL1	3.1
Exserohilum turcica Et28A	3.4
Fonsecaea erecta	2.8
Fonsecaea monophora	3.4
Fonsecaea nubica	3.2
Fusarium fujikuroi IMI 58289	4.3
Gloeophyllum trabeum ATCC 11539	2.9
Hyphopichia burtonii NRRL Y-1933	1.7
Kazachstania naganishii CBS 8797	1.8
Kluyveromyces marxianus DMKU3–1042	1.0
Kockovaella imperatae	1.9
Kwoniella bestiolae CBS 10118	2.6
Kwoniella dejecticola CBS 10117	2.8
Kwoniella mangroviensis CBS 8507	2.3
Kwoniella pini CBS 10737	2.0
Lobosporangium transversale	3.5
Meliniomyces bicolor E	4.2
Metschnikowia bicuspidata var. bicuspidata NRRL YB-4993	1.5
Millerozyma farinosa CBS 7064	1.3
Neofusicoccum parvum UCRNP2	2.6
Ogataea polymorpha	1.1
Paraphaeosphaeria sporulosa	4.0
Penicilliopsis zonata CBS 506.65	2.6
Penicillium arizonense	3.0
Penicillium digitatum Pd1	2.5
Phaeoacremonium minimum UCRPA7	2.3
Phialocephala scopiformis	4.1
Pichia kudriavzevii	1.5
Pichia membranifaciens NRRL Y-2026	1.2
Pneumocystis carinii B80	1.2
Pneumocystis jirovecii RU7	1.6
Pneumocystis murina B123	1.3
Pochonia chlamydosporia 170	4.0
Postia placenta MAD-698-R-SB12	3.0
Pseudocercospora fijiensis CIRAD86	3.0
Pseudogymnoascus destructans	2.5

(*Continued*)

Known as:	Florida Fungi Average
Pseudogymnoascus destructans 20631-21	2.6
Pseudogymnoascus verrucosus	3.2
Pseudozyma hubeiensis SY62	2.0
Purpureocillium lilacinum	2.8
Ramularia collo-cygni	3.0
Rhizopus microsporus ATCC 52813	1.9
Rhodotorula graminis WP1	3.3
Rhodotorula toruloides NP11	3.4
Saccharomyces eubayanus	1.4
Saitoella complicata NRRL Y-17804	2.0
Sclerotinia sclerotiorum 1980 UF-70	3.0
Sphaerulina musiva SO2202	2.5
Sugiyamaella lignohabitans	1.5
Suhomyces tanzawaensis NRRL Y-17324	1.4
Talaromyces atroroseus	3.8
Tetrapisispora blattae CBS 6284	1.6
Trichoderma asperellum CBS 433.97	3.5
Trichoderma citrinoviride	3.3
Trichoderma gamsii	3.6
Trichoderma harzianum CBS 226.95	3.5
Trichosporon asahii var. asahii CBS 2479	2.5
Tsuchiyaea wingfieldii CBS 7118	2.9
Wallemia ichthyophaga EXF-994	1.4
Wickerhamiella sorbophila	1.3
Wickerhamomyces anomalus NRRL Y-366-8	1.5
Xylona heveae TC161	2.1

Appendix J – Idaho Region Average for Fungi

Listing of Fungi detected and identified along with average number of unique peptides.

Known as:	Idaho Fungi Average
[Candida] pseudohaemulonis	5.7
Agaricus bisporus var. bisporus H97	7.7
Alternaria alternata	11.0
Amorphotheca resinae ATCC 22711	9.7
Ascoidea rubescens DSM 1968	3.7
Aspergillus aculeatus ATCC 16872	12.0
Aspergillus bombycis	11.3
Aspergillus campestris IBT 28561	8.7
Aspergillus candidus	7.3
Aspergillus glaucus CBS 516.65	11.0
Aspergillus novofumigatus IBT 16806	10.0
Aspergillus oryzae RIB40	6.7
Aspergillus steynii IBT 23096	10.7
Babjeviella inositovora NRRL Y-12698	3.0
Baudoinia panamericana UAMH 10762	11.3
Bipolaris maydis ATCC 48331	10.3
Bipolaris sorokiniana ND90Pr	10.3
Botrytis cinerea B05.10	10.0
Candida albicans SC5314	4.0
Candida tropicalis MYA-3404	3.3
Cercospora beticola	13.7
Coccidioides posadasii C735 delta SOWgp	7.3
Colletotrichum gloeosporioides Nara gc5	10.7
Colletotrichum higginsianum IMI 349063	13.7
Colletotrichum orchidophilum	10.7
Cordyceps fumosorosea ARSEF 2679	8.3
Cryptococcus amylolentus CBS 6039	7.7

(*Continued*)

Known as:	Idaho Fungi Average
Cyberlindnera jadinii NRRL Y-1542	5.3
Debaryomyces fabryi	6.0
Diplodia corticola	10.0
Eremothecium sinecaudum	4.0
Eutypa lata UCREL1	11.3
Exserohilum turcica Et28A	12.7
Fonsecaea erecta	9.7
Fonsecaea monophora	10.7
Fonsecaea nubica	9.3
Fusarium fujikuroi IMI 58289	11.7
Gloeophyllum trabeum ATCC 11539	8.7
Hyphopichia burtonii NRRL Y-1933	5.7
Kazachstania naganishii CBS 8797	6.0
Kluyveromyces marxianus DMKU3–1042	2.7
Kockovaella imperatae	5.3
Kwoniella bestiolae CBS 10118	9.7
Kwoniella dejecticola CBS 10117	7.0
Kwoniella mangroviensis CBS 8507	8.3
Kwoniella pini CBS 10737	7.3
Lobosporangium transversale	8.7
Meliniomyces bicolor E	13.0
Metschnikowia bicuspidata var. bicuspidata NRRL YB-4993	3.7
Millerozyma farinosa CBS 7064	5.0
Neofusicoccum parvum UCRNP2	9.0
Ogataea polymorpha	4.0
Paraphaeosphaeria sporulosa	13.7
Penicilliopsis zonata CBS 506.65	9.0
Penicillium arizonense	13.7
Penicillium digitatum Pd1	10.3
Phaeoacremonium minimum UCRPA7	10.7
Phialocephala scopiformis	15.0
Phycomycees blakesleeanus NRRL 1555(-)	7.3
Pichia kudriavzevii	5.0
Pichia membranifaciens NRRL Y-2026	6.0
Pneumocystis carinii B80	4.0
Pneumocystis jirovecii RU7	2.3
Pneumocystis murina B123	4.3
Pochonia chlamydosporia 170	14.3
Postia placenta MAD-698-R-SB12	8.3
Pseudocercospora fijiensis CIRAD86	11.3

(*Continued*)

Known as:	Idaho Fungi Average
Pseudogymnoascus destructans	9.3
Pseudogymnoascus destructans 20631-21	8.7
Pseudogymnoascus verrucosus	7.7
Pseudozyma hubeiensis SY62	8.0
Purpureocillium lilacinum	13.0
Ramularia collo-cygni	7.3
Rhizopus microsporus ATCC 52813	6.0
Rhodotorula graminis WP1	6.0
Rhodotorula toruloides NP11	11.0
Saccharomyces eubayanus	4.0
Saitoella complicata NRRL Y-17804	7.0
Sclerotinia sclerotiorum 1980 UF-70	10.0
Sphaerulina musiva SO2202	10.0
Sugiyamaella lignohabitans	7.0
Suhomyces tanzawaensis NRRL Y-17324	7.7
Talaromyces atroroseus	9.0
Tetrapisispora blattae CBS 6284	5.0
Trichoderma asperellum CBS 433.97	7.0
Trichoderma citrinoviride	8.0
Trichoderma gamsii	8.3
Trichoderma harzianum CBS 226.95	10.7
Trichosporon asahii var. asahii CBS 2479	5.7
Tsuchiyaea wingfieldii CBS 7118	7.7
Wallemia ichthyophaga EXF-994	5.0
Wickerhamiella sorbophila	1.7
Wickerhamomyces anomalus NRRL Y-366-8	5.3
Xylona heveae TC161	8.0

Appendix K – Iowa Region Average for Fungi

Listing of fungi detected and identified along with average number of unique peptides.

Known as:	Iowa Fungi Average
[Candida] pseudohaemulonis	7.0
Agaricus bisporus var. bisporus H97	10.6
Alternaria alternata	15.2
Amorphotheca resinae ATCC 22711	11.2
Ascoidea rubescens DSM 1968	6.0
Aspergillus aculeatus ATCC 16872	12.3
Aspergillus bombycis	13.5
Aspergillus campestris IBT 28561	10.4
Aspergillus candidus	10.5
Aspergillus glaucus CBS 516.65	10.0
Aspergillus novofumigatus IBT 16806	11.0
Aspergillus oryzae RIB40	12.2
Aspergillus steynii IBT 23096	14.1
Babjeviella inositovora NRRL Y-12698	6.8
Baudoinia panamericana UAMH 10762	9.9
Bipolaris maydis ATCC 48331	13.1
Bipolaris sorokiniana ND90Pr	13.0
Botrytis cinerea B05.10	13.7
Candida albicans SC5314	6.4
Candida tropicalis MYA-3404	7.1
Cercospora beticola	13.5
Coccidioides posadasii C735 delta SOWgp	9.0
Colletotrichum gloeosporioides Nara gc5	16.2
Colletotrichum higginsianum IMI 349063	16.7
Colletotrichum orchidophilum	14.5
Cordyceps fumosorosea ARSEF 2679	12.4
Cryptococcus amylolentus CBS 6039	10.3

(*Continued*)

Known as:	**Iowa Fungi Average**
Cyberlindnera jadinii NRRL Y-1542	6.1
Debaryomyces fabryi	6.1
Diplodia corticola	14.2
Eremothecium sinecaudum	4.7
Eutypa lata UCREL1	11.1
Exserohilum turcica Et28A	14.8
Fonsecaea erecta	13.8
Fonsecaea monophora	13.3
Fonsecaea nubica	13.5
Fusarium fujikuroi IMI 58289	15.1
Gloeophyllum trabeum ATCC 11539	11.7
Hyphopichia burtonii NRRL Y-1933	7.0
Kazachstania naganishii CBS 8797	6.6
Kluyveromyces marxianus DMKU3–1042	5.6
Kockovaella imperatae	9.1
Kwoniella bestiolae CBS 10118	11.6
Kwoniella dejecticola CBS 10117	11.4
Kwoniella mangroviensis CBS 8507	10.7
Kwoniella pini CBS 10737	10.2
Lobosporangium transversale	12.3
Meliniomyces bicolor E	17.4
Metschnikowia bicuspidata var. bicuspidata NRRL YB-4993	5.5
Millerozyma farinosa CBS 7064	9.9
Neofusicoccum parvum UCRNP2	10.1
Ogataea polymorpha	6.2
Paraphaeosphaeria sporulosa	14.4
Penicilliopsis zonata CBS 506.65	10.9
Penicillium arizonense	13.1
Penicillium digitatum Pd1	9.4
Phaeoacremonium minimum UCRPA7	9.2
Phialocephala scopiformis	17.3
Phycomycees blakesleeanus NRRL 1555(-)	11.3
Pichia kudriavzevii	5.6
Pichia membranifaciens NRRL Y-2026	6.4
Pneumocystis carinii B80	4.1
Pneumocystis jirovecii RU7	4.5
Pneumocystis murina B123	4.8
Pochonia chlamydosporia 170	15.2
Postia placenta MAD-698-R-SB12	12.2
Pseudocercospora fijiensis CIRAD86	14.0

(*Continued*)

Known as:	Iowa Fungi Average
Pseudogymnoascus destructans	10.2
Pseudogymnoascus destructans 20631-21	10.2
Pseudogymnoascus verrucosus	12.6
Pseudozyma hubeiensis SY62	10.7
Purpureocillium lilacinum	14.1
Ramularia collo-cygni	12.1
Rhizopus microsporus ATCC 52813	9.1
Rhodotorula graminis WP1	10.1
Rhodotorula toruloides NP11	11.0
Saccharomyces eubayanus	6.4
Saitoella complicata NRRL Y-17804	6.6
Sclerotinia sclerotiorum 1980 UF-70	11.9
Sphaerulina musiva SO2202	11.5
Sugiyamaella lignohabitans	6.0
Suhomyces tanzawaensis NRRL Y-17324	5.7
Talaromyces atroroseus	10.8
Tetrapisispora blattae CBS 6284	7.0
Trichoderma asperellum CBS 433.97	13.2
Trichoderma citrinoviride	11.4
Trichoderma gamsii	11.5
Trichoderma harzianum CBS 226.95	14.8
Trichosporon asahii var. asahii CBS 2479	8.6
Tsuchiyaea wingfieldii CBS 7118	9.6
Wallemia ichthyophaga EXF-994	6.0
Wickerhamiella sorbophila	4.4
Wickerhamomyces anomalus NRRL Y-366-8	6.4
Xylona heveae TC161	9.4

Appendix L – Montana Region Average for Fungi

Listing of fungi detected and identified along with average number of unique peptides.

Known as:	Montana Fungi Average
[Candida] pseudohaemulonis	4.5
Agaricus bisporus var. bisporus H97	12
Alternaria alternata	19
Amorphotheca resinae ATCC 22711	11.5
Ascoidea rubescens DSM 1968	9.5
Aspergillus aculeatus ATCC 16872	12
Aspergillus bombycis	12.5
Aspergillus campestris IBT 28561	11.8
Aspergillus candidus	10.3
Aspergillus glaucus CBS 516.65	10.3
Aspergillus novofumigatus IBT 16806	11.5
Aspergillus oryzae RIB40	14
Aspergillus steynii IBT 23096	13.3
Babjeviella inositovora NRRL Y-12698	8.8
Baudoinia panamericana UAMH 10762	11.3
Bipolaris maydis ATCC 48331	13
Bipolaris sorokiniana ND90Pr	11.5
Botrytis cinerea B05.10	11.8
Candida albicans SC5314	5.8
Candida tropicalis MYA-3404	3.8
Cercospora beticola	15.3
Coccidioides posadasii C735 delta SOWgp	10.8
Colletotrichum gloeosporioides Nara gc5	17.3
Colletotrichum higginsianum IMI 349063	20.3
Colletotrichum orchidophilum	15.3
Cordyceps fumosorosea ARSEF 2679	12.5

(*Continued*)

Known as:	Montana Fungi Average
Cryptococcus amylolentus CBS 6039	11
Cyberlindnera jadinii NRRL Y-1542	3.5
Debaryomyces fabryi	9
Diplodia corticola	12
Eremothecium sinecaudum	6.3
Eutypa lata UCREL1	10.5
Exserohilum turcica Et28A	10.5
Fonsecaea erecta	13.3
Fonsecaea monophora	16.8
Fonsecaea nubica	15.5
Fusarium fujikuroi IMI 58289	19.8
Gloeophyllum trabeum ATCC 11539	9.8
Hyphopichia burtonii NRRL Y-1933	7.3
Kazachstania naganishii CBS 8797	4.8
Kluyveromyces marxianus DMKU3–1042	5.8
Kockovaella imperatae	8.3
Kwoniella bestiolae CBS 10118	13.3
Kwoniella dejecticola CBS 10117	14.3
Kwoniella mangroviensis CBS 8507	9.8
Kwoniella pini CBS 10737	12.5
Lobosporangium transversale	10.8
Meliniomyces bicolor E	16.3
Metschnikowia bicuspidata var. bicuspidata NRRL YB-4993	6.3
Millerozyma farinosa CBS 7064	9.8
Neofusicoccum parvum UCRNP2	8.5
Ogataea polymorpha	8
Paraphaeosphaeria sporulosa	14.5
Penicilliopsis zonata CBS 506.65	10.3
Penicillium arizonense	11.8
Penicillium digitatum Pd1	7.3
Phaeoacremonium minimum UCRPA7	12
Phialocephala scopiformis	19.3
Phycomyces blakesleeanus NRRL 1555(-)	12.3
Pichia kudriavzevii	7
Pichia membranifaciens NRRL Y-2026	6.5
Pneumocystis carinii B80	2.3
Pneumocystis jirovecii RU7	4.5
Pneumocystis murina B123	3
Pochonia chlamydosporia 170	15.5
Postia placenta MAD-698-R-SB12	11.3
Pseudocercospora fijiensis CIRAD86	13.5

(*Continued*)

Known as:	Montana Fungi Average
Pseudogymnoascus destructans	8.8
Pseudogymnoascus destructans 20631-21	9
Pseudogymnoascus verrucosus	11
Pseudozyma hubeiensis SY62	9
Purpureocillium lilacinum	17.8
Ramularia collo-cygni	15.5
Rhizopus microsporus ATCC 52813	13.5
Rhodotorula graminis WP1	11.3
Rhodotorula toruloides NP11	12.3
Saccharomyces eubayanus	6
Saitoella complicata NRRL Y-17804	5.3
Sclerotinia sclerotiorum 1980 UF-70	13.5
Sphaerulina musiva SO2202	13.3
Sugiyamaella lignohabitans	6.8
Suhomyces tanzawaensis NRRL Y-17324	6.3
Talaromyces atroroseus	14.3
Tetrapisispora blattae CBS 6284	6.3
Trichoderma asperellum CBS 433.97	14.8
Trichoderma citrinoviride	12.3
Trichoderma gamsii	14.5
Trichoderma harzianum CBS 226.95	16.3
Trichosporon asahii var. asahii CBS 2479	10
Tsuchiyaea wingfieldii CBS 7118	9.8
Wallemia ichthyophaga EXF-994	8
Wickerhamiella sorbophila	4.3
Wickerhamomyces anomalus NRRL Y-366-8	6
Xylona heveae TC161	10.8

Appendix M – National Average for Viruses

Listing of viruses detected and identified along with average number of unique peptides.

Known as:	National Average for Viruses
a A virus (A/Goose/Guangdong/1/96(H5N1))	1.9
a A virus (A/Hong Kong/1073/99(H9N2))	1.9
a A virus (A/Korea/426/1968(H2N2))	1.6
a A virus (A/New York/392/2004(H3N2))	1.3
a A virus (A/Puerto Rico/8/1934(H1N1))	0.0
a C virus (C/Ann Arbor/1/50)	1.8
African horse sickness virus	2.2
African swine fever virus	16.4
Akabane virus	1.1
Alcelaphine herpesvirus 1	8.6
Alkhumra hemorrhagic fever virus	1.1
Allpahuayo virus	1.0
Amapari virus	1.1
Andes virus	1.1
Banana bunchy top virus	0.4
Barley yellow dwarf virus-GAV	0.7
Barley yellow dwarf virus-MAV	0.5
Barley yellow dwarf virus-PAS	0.4
Barley yellow dwarf virus-PAV	0.9
Bear Canyon virus	1.2
Bluetongue virus	2.0
Bovine ephemeral fever virus	1.2
Bundibugyo ebolavirus	1.4
Camelpox virus	18.1
Cercopithecine herpesvirus 2	13.5
Cercopithecine herpesvirus 5	15.9
Cercopithecine herpesvirus 9	10.8

(Continued)

Known as:	National Average for Viruses
Chapare virus	0.9
Chikungunya virus	1.3
Classical swine fever virus	1.4
Colorado tick fever virus	3.2
Crimean-Congo hemorrhagic fever virus	2.4
Cupixi virus	0.9
Dengue virus 1	1.0
Dengue virus 2	1.3
Dengue virus 3	1.0
Dengue virus 4	1.3
Dobrava-Belgrade virus	0.6
Eastern equine encephalitis virus	1.2
Foot-and-mouth disease virus - type A	0.7
Foot-and-mouth disease virus - type Asia 1	0.6
Foot-and-mouth disease virus - type C	0.7
Foot-and-mouth disease virus - type O	0.6
Foot-and-mouth disease virus - type SAT 1	0.6
Foot-and-mouth disease virus - type SAT 2	0.5
Foot-and-mouth disease virus - type SAT 3	0.5
Goatpox virus Pellor	15.8
Hantaan virus	1.4
Hantavirus Z10	1.1
Hendra virus	2.2
Influenza B virus	1.8
Japanese encephalitis virus	1.0
Junin virus	1.1
Karshi virus	0.9
Langat virus	1.4
Lassa virus	1.2
Latino virus	1.2
Louping ill virus	1.6
Lujo virus	1.6
Lumpy skin disease virus NI-2490	14.5
Lymphocytic choriomeningitis virus	1.4
Machupo virus	1.1
Marburg marburgvirus	1.4
Menangle virus	1.2
Monkeypox virus Zaire-96-I-16	18.1
Newcastle disease virus B1	1.4
Nipah virus	2.1
Oliveros virus	0.8
Omsk hemorrhagic fever virus	1.3

(Continued)

Known as:	National Average for Viruses
Parana virus	1.4
Peste-des-petits-ruminants virus	2.0
Pichinde virus	0.9
Porcine teschovirus	0.6
Powassan virus	1.4
Puumala virus	0.4
Reston ebolavirus	1.2
Rift Valley fever virus	0.8
Sabia virus	1.3
Sandfly fever Naples virus	1.3
Seoul virus	0.7
Severe acute respiratory syndrome-related coronavirus	0.0
Sheeppox virus	15.0
Simian hemorrhagic fever virus	1.3
Sin Nombre virus	1.1
st virus (strain Kabete O)	1.4
Sudan ebolavirus	1.9
Suid herpesvirus 1	11.6
Swinepox virus	14.3
Tacaribe virus	1.4
Tai Forest ebolavirus	1.5
Tamiami virus	1.0
Thottapalayam virus	1.1
Tick-borne encephalitis virus	1.4
Tula virus	0.5
Vaccinia virus	17.4
Variola virus	17.9
Venezuelan equine encephalitis virus	1.8
Vesicular exanthema of swine virus	0.9
Vesicular stomatitis Indiana virus	1.4
Western equine encephalitis virus	1.4
Whitewater Arroyo virus	1.2
Yeast alcohol dehydrogenase 1	0.3
Yellow fever virus	1.0
Zaire ebola virus	1.2

Appendix N – California Region Average for Viruses

Listing of viruses detected and identified along with average number of unique peptides.

Known as:	California Virus Average
a A virus (A/Goose/Guangdong/1/96(H5N1))	1.6
a A virus (A/Hong Kong/1073/99(H9N2))	0.9
a A virus (A/Korea/426/1968(H2N2))	1.6
a A virus (A/New York/392/2004(H3N2))	2.3
a A virus (A/Puerto Rico/8/1934(H1N1))	0.0
a C virus (C/Ann Arbor/1/50)	1.9
African horse sickness virus	2.2
African swine fever virus	18.4
Akabane virus	1.4
Alcelaphine herpesvirus 1	9.1
Alkhumra hemorrhagic fever virus	1.3
Allpahuayo virus	1.4
Amapari virus	1.4
Andes virus	0.6
Banana bunchy top virus	0.5
Barley yellow dwarf virus-GAV	0.6
Barley yellow dwarf virus-MAV	0.6
Barley yellow dwarf virus-PAS	0.9
Barley yellow dwarf virus-PAV	0.7
Bear Canyon virus	1.9
Bluetongue virus	4.1
Bovine ephemeral fever virus	1.2
Bundibugyo ebolavirus	2.5
Camelpox virus	20.6
Cercopithecine herpesvirus 2	14.5
Cercopithecine herpesvirus 5	14.9
Cercopithecine herpesvirus 9	10.5
Chapare virus	1.4
Chikungunya virus	1.0

(*Continued*)

Known as:	California Virus Average
Classical swine fever virus	2.4
Colorado tick fever virus	3.4
Crimean-Congo hemorrhagic fever virus	2.0
Cupixi virus	1.1
Dengue virus 1	1.4
Dengue virus 2	1.7
Dengue virus 3	1.2
Dengue virus 4	1.5
Dobrava-Belgrade virus	1.2
Eastern equine encephalitis virus	1.2
Foot-and-mouth disease virus - type A	1.1
Foot-and-mouth disease virus - type Asia 1	1.1
Foot-and-mouth disease virus - type C	0.9
Foot-and-mouth disease virus - type O	1.0
Foot-and-mouth disease virus - type SAT 1	0.8
Foot-and-mouth disease virus - type SAT 2	0.7
Foot-and-mouth disease virus - type SAT 3	0.9
Goatpox virus Pellor	19.9
Hantaan virus	0.9
Hantavirus Z10	0.8
Hendra virus	1.9
Influenza B virus	2.1
Japanese encephalitis virus	1.4
Junin virus	1.0
Karshi virus	1.4
Langat virus	1.4
Lassa virus	0.8
Latino virus	1.3
Louping ill virus	1.3
Lujo virus	1.7
Lumpy skin disease virus NI-2490	19.0
Lymphocytic choriomeningitis virus	1.1
Machupo virus	1.2
Marburg marburgvirus	1.6
Menangle virus	1.1
Monkeypox virus Zaire-96-I-16	22.6
Newcastle disease virus B1	1.4
Nipah virus	1.8
Oliveros virus	1.5
Omsk hemorrhagic fever virus	1.4
Parana virus	0.9
Peste-des-petits-ruminants virus	2.4
Pichinde virus	1.4

(*Continued*)

Known as:	California Virus Average
Porcine teschovirus	0.5
Powassan virus	1.3
Puumala virus	0.4
Reston ebolavirus	1.1
Rift Valley fever virus	1.4
Sabia virus	1.1
Sandfly fever Naples virus	1.0
Seoul virus	1.1
Severe acute respiratory syndrome-related coronavirus	0.1
Sheeppox virus	18.3
Simian hemorrhagic fever virus	1.9
Sin Nombre virus	0.9
st virus (strain Kabete O)	2.6
Sudan ebolavirus	1.8
Suid herpesvirus 1	14.0
Swinepox virus	17.8
Tacaribe virus	1.6
Tai Forest ebolavirus	1.8
Tamiami virus	1.4
Thottapalayam virus	1.0
Tick-borne encephalitis virus	1.6
Tula virus	1.4
Vaccinia virus	20.4
Variola virus	22.1
Venezuelan equine encephalitis virus	1.4
Vesicular exanthema of swine virus	1.1
Vesicular stomatitis Indiana virus	1.8
Western equine encephalitis virus	1.3
Whitewater Arroyo virus	1.3
Yeast alcohol dehydrogenase 1	0.3
Yellow fever virus	1.5
Zaire ebola virus	1.4

Appendix O – Florida Region Average for Viruses

Listing of viruses detected and identified along with average number of unique peptides.

Known as:	Florida Virus Average
a A virus (A/Goose/Guangdong/1/96(H5N1))	1.2
a A virus (A/Hong Kong/1073/99(H9N2))	1.1
a A virus (A/Korea/426/1968(H2N2))	1.7
a A virus (A/New York/392/2004(H3N2))	1.6
a A virus (A/Puerto Rico/8/1934(H1N1))	0.0
a C virus (C/Ann Arbor/1/50)	1.4
African horse sickness virus	1.4
African swine fever virus	12.4
Akabane virus	0.8
Alcelaphine herpesvirus 1	5.7
Alkhumra hemorrhagic fever virus	0.6
Allpahuayo virus	0.6
Amapari virus	0.9
Andes virus	1.1
Banana bunchy top virus	0.2
Barley yellow dwarf virus-GAV	0.4
Barley yellow dwarf virus-MAV	0.3
Barley yellow dwarf virus-PAS	0.4
Barley yellow dwarf virus-PAV	0.4
Bear Canyon virus	1.0
Bluetongue virus	1.6
Bovine ephemeral fever virus	1.0
Bundibugyo ebolavirus	1.0
Camelpox virus	12.4
Cercopithecine herpesvirus 2	9.9
Cercopithecine herpesvirus 5	10.6

(*Continued*)

Known as:	Florida Virus Average
Cercopithecine herpesvirus 9	6.0
Chapare virus	1.1
Chikungunya virus	0.4
Classical swine fever virus	1.1
Colorado tick fever virus	2.0
Crimean-Congo hemorrhagic fever virus	1.5
Cupixi virus	0.7
Dengue virus 1	0.8
Dengue virus 2	0.7
Dengue virus 3	0.7
Dengue virus 4	0.9
Dobrava-Belgrade virus	1.0
Eastern equine encephalitis virus	0.4
Foot-and-mouth disease virus - type A	0.2
Foot-and-mouth disease virus - type Asia 1	0.3
Foot-and-mouth disease virus - type C	0.2
Foot-and-mouth disease virus - type O	0.4
Foot-and-mouth disease virus - type SAT 1	0.7
Foot-and-mouth disease virus - type SAT 2	0.4
Foot-and-mouth disease virus - type SAT 3	0.8
Goatpox virus Pellor	12.3
Hantaan virus	1.4
Hantavirus Z10	1.5
Hendra virus	1.5
Influenza B virus	1.4
Japanese encephalitis virus	0.7
Junin virus	0.8
Karshi virus	0.8
Langat virus	0.8
Lassa virus	1.0
Latino virus	1.1
Louping ill virus	0.9
Lujo virus	0.9
Lumpy skin disease virus NI-2490	12.3
Lymphocytic choriomeningitis virus	0.6
Machupo virus	1.0
Marburg marburgvirus	1.0
Menangle virus	1.0

(Continued)

Known as:	Florida Virus Average
Monkeypox virus Zaire-96-I-16	12.3
Newcastle disease virus B1	1.3
Nipah virus	0.9
Oliveros virus	0.4
Omsk hemorrhagic fever virus	0.5
Parana virus	0.8
Peste-des-petits-ruminants virus	1.4
Pichinde virus	1.0
Porcine teschovirus	0.4
Powassan virus	1.1
Puumala virus	0.4
Reston ebolavirus	1.0
Rift Valley fever virus	1.4
Sabia virus	0.9
Sandfly fever Naples virus	1.0
Seoul virus	1.2
Severe acute respiratory syndrome-related coronavirus	0.2
Sheeppox virus	12.1
Simian hemorrhagic fever virus	0.9
Sin Nombre virus	1.1
st virus (strain Kabete O)	1.6
Sudan ebolavirus	1.4
Suid herpesvirus 1	6.5
Swinepox virus	10.6
Tacaribe virus	0.7
Tai Forest ebolavirus	0.8
Tamiami virus	0.9
Thottapalayam virus	0.7
Tick-borne encephalitis virus	0.5
Tula virus	1.1
Vaccinia virus	12.6
Variola virus	11.5
Venezuelan equine encephalitis virus	1.2
Vesicular exanthema of swine virus	0.3
Vesicular stomatitis Indiana virus	1.0
Western equine encephalitis virus	1.2
Whitewater Arroyo virus	0.7
Yeast alcohol dehydrogenase 1	0.1
Yellow fever virus	0.8
Zaire ebola virus	1.3

Appendix P – Idaho Region Average for Viruses

Listing of viruses detected and identified along with average number of unique peptides.

Known as:	Idaho Virus Average
a A virus (A/Goose/Guangdong/1/96(H5N1))	2.5
a A virus (A/Hong Kong/1073/99(H9N2))	1.5
a A virus (A/Korea/426/1968(H2N2))	2.0
a A virus (A/New York/392/2004(H3N2))	1.5
a A virus (A/Puerto Rico/8/1934(H1N1))	0.0
a C virus (C/Ann Arbor/1/50)	3.3
African horse sickness virus	1.3
African swine fever virus	19.3
Akabane virus	1.8
Alcelaphine herpesvirus 1	9.0
Alkhumra hemorrhagic fever virus	1.5
Allpahuayo virus	1.3
Amapari virus	0.8
Andes virus	1.3
Banana bunchy top virus	0.0
Barley yellow dwarf virus-GAV	2.0
Barley yellow dwarf virus-MAV	1.3
Barley yellow dwarf virus-PAS	0.3
Barley yellow dwarf virus-PAV	0.5
Bear Canyon virus	1.5
Bluetongue virus	3.0
Bovine ephemeral fever virus	2.5
Bundibugyo ebolavirus	1.5
Camelpox virus	22.8
Cercopithecine herpesvirus 2	14.8
Cercopithecine herpesvirus 5	15.3
Cercopithecine herpesvirus 9	10.3

(*Continued*)

Known as:	Idaho Virus Average
Chapare virus	1.3
Chikungunya virus	1.5
Classical swine fever virus	1.3
Colorado tick fever virus	2.8
Crimean-Congo hemorrhagic fever virus	1.8
Cupixi virus	0.5
Dengue virus 1	1.0
Dengue virus 2	2.0
Dengue virus 3	2.8
Dengue virus 4	2.5
Dobrava-Belgrade virus	1.3
Eastern equine encephalitis virus	2.0
Foot-and-mouth disease virus - type A	1.0
Foot-and-mouth disease virus - type Asia 1	2.3
Foot-and-mouth disease virus - type C	1.3
Foot-and-mouth disease virus - type O	0.8
Foot-and-mouth disease virus - type SAT 1	0.3
Foot-and-mouth disease virus - type SAT 2	0.8
Foot-and-mouth disease virus - type SAT 3	0.5
Goatpox virus Pellor	16.5
Hantaan virus	1.5
Hantavirus Z10	0.8
Hendra virus	1.0
Influenza B virus	1.8
Japanese encephalitis virus	1.3
Junin virus	1.3
Karshi virus	0.8
Langat virus	0.8
Lassa virus	0.5
Latino virus	1.0
Louping ill virus	1.0
Lujo virus	1.3
Lumpy skin disease virus NI-2490	16.0
Lymphocytic choriomeningitis virus	1.0
Machupo virus	0.8
Marburg marburgvirus	1.5
Menangle virus	1.3
Monkeypox virus Zaire-96-I-16	18.5

(Continued)

Known as:	Idaho Virus Average
Newcastle disease virus B1	2.0
Nipah virus	1.5
Oliveros virus	1.0
Omsk hemorrhagic fever virus	1.3
Parana virus	0.5
Peste-des-petits-ruminants virus	1.3
Pichinde virus	0.3
Porcine teschovirus	0.5
Powassan virus	1.8
Puumala virus	0.8
Reston ebolavirus	1.5
Rift Valley fever virus	1.8
Sabia virus	0.5
Sandfly fever Naples virus	2.0
Seoul virus	1.8
Severe acute respiratory syndrome-related coronavirus	0.0
Sheeppox virus	15.8
Simian hemorrhagic fever virus	1.0
Sin Nombre virus	1.5
st virus (strain Kabete O)	1.0
Sudan ebolavirus	1.3
Suid herpesvirus 1	9.5
Swinepox virus	14.8
Tacaribe virus	1.5
Tai Forest ebolavirus	0.5
Tamiami virus	1.8
Thottapalayam virus	1.0
Tick-borne encephalitis virus	0.8
Tula virus	1.3
Vaccinia virus	20.0
Variola virus	22.0
Venezuelan equine encephalitis virus	2.3
Vesicular exanthema of swine virus	0.5
Vesicular stomatitis Indiana virus	1.5
Western equine encephalitis virus	1.0
Whitewater Arroyo virus	1.8
Yeast alcohol dehydrogenase 1	0.0
Yellow fever virus	1.0
Zaire ebola virus	1.3

Appendix Q – Iowa Region Average for Viruses

Listing of viruses detected and identified along with average number of unique peptides.

Known as:	Iowa Virus Average
a A virus (A/Goose/Guangdong/1/96(H5N1))	4.6
a A virus (A/Hong Kong/1073/99(H9N2))	5.1
a A virus (A/Korea/426/1968(H2N2))	4.8
a A virus (A/New York/392/2004(H3N2))	4.8
a A virus (A/Puerto Rico/8/1934(H1N1))	0.0
a C virus (C/Ann Arbor/1/50)	4.4
African horse sickness virus	7.1
African swine fever virus	44.5
Akabane virus	3.0
Alcelaphine herpesvirus 1	23.3
Alkhumra hemorrhagic fever virus	2.7
Allpahuayo virus	3.4
Amapari virus	2.5
Andes virus	3.5
Banana bunchy top virus	0.6
Barley yellow dwarf virus-GAV	1.6
Barley yellow dwarf virus-MAV	1.7
Barley yellow dwarf virus-PAS	1.2
Barley yellow dwarf virus-PAV	1.6
Bear Canyon virus	3.4
Bluetongue virus	7.1
Bovine ephemeral fever virus	5.1
Bundibugyo ebolavirus	4.3
Camelpox virus	48.4
Cercopithecine herpesvirus 2	37.2
Cercopithecine herpesvirus 5	42.8
Cercopithecine herpesvirus 9	25.8

(*Continued*)

Known as:	Iowa Virus Average
Chapare virus	4.0
Chikungunya virus	3.5
Classical swine fever virus	3.7
Colorado tick fever virus	7.5
Crimean-Congo hemorrhagic fever virus	7.0
Cupixi virus	3.4
Dengue virus 1	2.8
Dengue virus 2	3.7
Dengue virus 3	2.6
Dengue virus 4	2.6
Dobrava-Belgrade virus	2.3
Eastern equine encephalitis virus	3.6
Foot-and-mouth disease virus - type A	2.4
Foot-and-mouth disease virus - type Asia 1	2.2
Foot-and-mouth disease virus - type C	2.0
Foot-and-mouth disease virus - type O	2.1
Foot-and-mouth disease virus - type SAT 1	2.1
Foot-and-mouth disease virus - type SAT 2	1.9
Foot-and-mouth disease virus - type SAT 3	2.2
Goatpox virus Pellor	45.0
Hantaan virus	3.0
Hantavirus Z10	3.1
Hendra virus	4.7
Influenza B virus	4.5
Japanese encephalitis virus	3.4
Junin virus	3.5
Karshi virus	3.3
Langat virus	3.3
Lassa virus	2.4
Latino virus	3.3
Louping ill virus	3.2
Lujo virus	3.1
Lumpy skin disease virus NI-2490	46.5
Lymphocytic choriomeningitis virus	3.6
Machupo virus	3.4
Marburg marburgvirus	2.9
Menangle virus	2.5
Monkeypox virus Zaire-96-I-16	46.9

(Continued)

Known as:	Iowa Virus Average
Newcastle disease virus B1	2.9
Nipah virus	4.7
Oliveros virus	2.4
Omsk hemorrhagic fever virus	3.1
Parana virus	2.0
Peste-des-petits-ruminants virus	4.3
Pichinde virus	3.2
Porcine teschovirus	2.0
Powassan virus	3.4
Puumala virus	1.2
Reston ebolavirus	3.2
Rift Valley fever virus	3.8
Sabia virus	3.5
Sandfly fever Naples virus	3.8
Seoul virus	2.7
Severe acute respiratory syndrome-related coronavirus	0.2
Sheeppox virus	44.4
Simian hemorrhagic fever virus	4.3
Sin Nombre virus	3.6
st virus (strain Kabete O)	4.5
Sudan ebolavirus	3.6
Suid herpesvirus 1	29.4
Swinepox virus	42.4
Tacaribe virus	3.0
Tai Forest ebolavirus	3.5
Tamiami virus	3.2
Thottapalayam virus	3.5
Tick-borne encephalitis virus	3.0
Tula virus	3.0
Vaccinia virus	49.2
Variola virus	48.9
Venezuelan equine encephalitis virus	4.6
Vesicular exanthema of swine virus	2.7
Vesicular stomatitis Indiana virus	3.5
Western equine encephalitis virus	3.7
Whitewater Arroyo virus	3.6
Yeast alcohol dehydrogenase 1	0.2
Yellow fever virus	2.9
Zaire ebola virus	3.4

Appendix R – Montana Region Average for Viruses

Listing of viruses detected and identified along with average number of unique peptides.

Known as:	Montana Virus Average
a A virus (A/Goose/Guangdong/1/96(H5N1))	5.3
a A virus (A/Hong Kong/1073/99(H9N2))	5.3
a A virus (A/Korea/426/1968(H2N2))	6.0
a A virus (A/New York/392/2004(H3N2))	4.5
a A virus (A/Puerto Rico/8/1934(H1N1))	0.0
a C virus (C/Ann Arbor/1/50)	5.3
African horse sickness virus	8.5
African swine fever virus	62.3
Akabane virus	5.8
Alcelaphine herpesvirus 1	32.5
Alkhumra hemorrhagic fever virus	3.5
Allpahuayo virus	4.5
Amapari virus	3.0
Andes virus	3.5
Banana bunchy top virus	0.5
Barley yellow dwarf virus-GAV	2.8
Barley yellow dwarf virus-MAV	2.0
Barley yellow dwarf virus-PAS	2.5
Barley yellow dwarf virus-PAV	2.8
Bear Canyon virus	5.3
Bluetongue virus	8.8
Bovine ephemeral fever virus	9.0
Bundibugyo ebolavirus	6.0
Camelpox virus	85.8
Cercopithecine herpesvirus 2	51.0
Cercopithecine herpesvirus 5	64.0
Cercopithecine herpesvirus 9	30.8

(Continued)

Known as:	Montana Virus Average
Chapare virus	5.5
Chikungunya virus	5.8
Classical swine fever virus	5.0
Colorado tick fever virus	12.8
Crimean-Congo hemorrhagic fever virus	7.8
Cupixi virus	3.0
Dengue virus 1	4.8
Dengue virus 2	5.8
Dengue virus 3	3.5
Dengue virus 4	5.0
Dobrava-Belgrade virus	4.0
Eastern equine encephalitis virus	5.5
Foot-and-mouth disease virus - type A	1.8
Foot-and-mouth disease virus - type Asia 1	3.3
Foot-and-mouth disease virus - type C	2.3
Foot-and-mouth disease virus - type O	1.8
Foot-and-mouth disease virus - type SAT 1	3.0
Foot-and-mouth disease virus - type SAT 2	2.8
Foot-and-mouth disease virus - type SAT 3	3.3
Goatpox virus Pellor	64.8
Hantaan virus	4.0
Hantavirus Z10	5.0
Hendra virus	7.8
Influenza B virus	6.3
Japanese encephalitis virus	4.5
Junin virus	5.0
Karshi virus	4.5
Langat virus	6.0
Lassa virus	6.0
Latino virus	4.0
Louping ill virus	4.5
Lujo virus	3.8
Lumpy skin disease virus NI-2490	70.0
Lymphocytic choriomeningitis virus	4.8
Machupo virus	5.0
Marburg marburgvirus	3.8
Menangle virus	5.0
Monkeypox virus Zaire-96-I-16	81.0

(Continued)

Known as:	Montana Virus Average
Newcastle disease virus B1	7.8
Nipah virus	9.3
Oliveros virus	3.5
Omsk hemorrhagic fever virus	5.3
Parana virus	4.5
Peste-des-petits-ruminants virus	6.5
Pichinde virus	5.5
Porcine teschovirus	3.3
Powassan virus	6.3
Puumala virus	1.8
Reston ebolavirus	5.3
Rift Valley fever virus	5.8
Sabia virus	3.8
Sandfly fever Naples virus	7.0
Seoul virus	4.0
Severe acute respiratory syndrome-related coronavirus	0.0
Sheeppox virus	62.0
Simian hemorrhagic fever virus	7.0
Sin Nombre virus	6.5
st virus (strain Kabete O)	5.5
Sudan ebolavirus	5.3
Suid herpesvirus 1	41.8
Swinepox virus	61.5
Tacaribe virus	4.8
Tai Forest ebolavirus	4.5
Tamiami virus	5.3
Thottapalayam virus	6.3
Tick-borne encephalitis virus	5.5
Tula virus	3.8
Vaccinia virus	88.5
Variola virus	82.0
Venezuelan equine encephalitis virus	5.3
Vesicular exanthema of swine virus	3.5
Vesicular stomatitis Indiana virus	2.5
Western equine encephalitis virus	5.8
Whitewater Arroyo virus	4.5
Yeast alcohol dehydrogenase 1	0.5
Yellow fever virus	3.3
Zaire ebola virus	5.3

Index

Note: *Italicized* page numbers refer to figures.